GROUND WATER POLLUTION

A Bibliography

W. K. Summers
Consulting Geologist
Socorro, New Mexico

and

Zane Spiegel
Consulting Geologist
Santa Fe, New Mexico

 **ann arbor science** PUBLISHERS INC.
POST OFFICE BOX 1425 • ANN ARBOR, MICHIGAN 48106

Second Printing 1975
Third Printing 1975

Copyright © 1974 by Ann Arbor Science Publishers, Inc.
P.O. Box 1425, Ann Arbor, Michigan 48106

Library of Congress Catalog Card No. 73-87165
ISBN 0-250-40048-0

PREFACE

This partially annotated bibliography is the product of a concentrated literature search particularly in the areas of ground water contamination by nitrates, heavy metals, pesticides and herbicides. The impact of urbanization and the effects of solid waste disposal, animal wastes and petroleum products on ground water quality are covered in the more than 400 entries.

W. K. Summers
Socorro, New Mexico
January 1974

SOURCES OF CITATIONS

The material presented was selected from the extensive files of the compiler and the following bibliographic sources:

LaMoreaux, P.E., Raymond, D., and Joiner, T.J. Hydrology of Limestone Terraines--Annotated Bibliography of Carbonate Rocks. Alabama Geological Survey, Bull. 94, Part A, 242 pp. (1970).

+ + +

Dugan, P., Pfister, R.M., and Spraque, M.L. Bibliography of Organic Pesticide Publications Having Relevance to Public Health and Water Pollution Problems. New York State Department of Health, Research Report No. 10, Part 2, 121 pp. (1963).

+ + +

Dugan, P., Pfister, R.M., and Spraque, M.L. Evaluation of the Extent and Nature of Pesticide and Detergent Involvement in Surface Waters of a Selected Watershed. New York State Department of Health, Research Report No. 10, Part 1, 74 pp. (1963).

+ + +

Bailey, G.W. Role of Soils and Sediments in Water Pollution Control. Part 1: Reactions of Nitrogenous and Phosphate Compounds with Soils and Geologic Stratum. U.S. Dept. of the Interior, Federal Water Pollution Control Administration, Southeast Laboratory, 90 pp. (1968).

Office of Water Resource Research. Soil Nitrogen Cycle, A Bibliography. Water Resource Scientific Information Center, WRSIC 72-208, 296 pp. (1972).

+ + +

Office of Water Resource Research. Arsenic and Lead in Water, A Bibliography. Water Resource Scientific Information Center, WRSIC 71-209 (1971).

+ + +

Office of Water Resource Research. Urbanization and Sedimentation, A Bibliography. Water Resource Scientific Information Center, WRSIC 71-203, 116 pp. (1971).

+ + +

MacNish, R.D., Heath, R.C., Johnson, L.E., Wilkens, R.A., and Duryea, R.D. Bibliography of the Ground Water Resources of New York through 1967. State of New York Conservation Department, Water Resources Commission, Bull. 66, 186 pp. (1969).

+ + +

Randolph, J.R., and Deike, R.G. Bibliography of Hydrology of the United States, 1963. USGS WSP 1863, 166 pp. (1966).

+ + +

Randolph, J.R., Baker, N.M., and Deike, R.G. Bibliography of Hydrology of the United States and Canada, 1964. USGS WSP 1864, 232 pp. (1969).

+ + +

Vorhis, R.C. Bibliography of Publications Relating to Ground Water Prepared by the Geological Survey and Cooperating Agencies 1946-1955. USGS WSP 1492, 203 pp. (1957).

+ + +

Waring, G.A., and Meinzer, O.E. Bibliography and Index of Publications Related to Ground Water Prepared by the Geological Survey and Cooperating Agencies. USGS WSP 992, 412 pp. (1947).

Office of Water Resource Research. <u>Subsurface Water Pollu-</u>
<u>tion: A Selective Annotated Bibliography</u>. (1972)

Part 1: Subsurface Wastes Disposal (156 pp.)
Part 2: Saline Water Intrusion (161 pp.)
Part 3: Percolation from Surface Sources (162 pp.)

+ + +

Thomas, R.E., Cohen, J.M., and Bendixen, T.W. <u>Pesticides</u>
<u>in Soil and Water, An Annotated Bibliography</u>. USDHEW,
PHS, Publication No. 999-WP-17 (1964).

+ + +

Viets, F.G., and Hageman, R.H. <u>Factors Affecting the</u>
<u>Accumulation of Nitrate in Soil, Water and Plants</u>.
USDA, ARS, Ag. Handbook No. 413, pp. 55-63 (1971).

+ + +

Department of Scientific and Industrial Research (Great
Britain). <u>Water Pollution Abstracts</u>, Volumes 38 to 49
(1965-1972).

+ + +

USDA. <u>Abstracts of Recently Published Material on Soil</u>
<u>and Water Concentration</u>. ARS, Nos. 9-28, 30-41.

+ + +

Abstracts of North American Geology (1969-1971).

+ + +

Bibliography and Index of Geology, Vol. 36 (1972).

ABBREVIATIONS

For the most part fairly standard or obvious abbreviations have been used in the citations. Notable exceptions include:

FWPCA: Federal Water Pollution Control Administration

NGWQS: National Ground Water Quality Symposium, U.S.
            Environmental Protection Agency, Water Pollution
            Research Series, No. 16060 GRB 08/71.

USDHEW   U.S. Department of Health, Education and Welfare,
PHS: Public Health Service

USGS WSP: U.S. Geological Survey, Water Supply Paper

WPCRS: Water Pollution Control Research Series

WRB: Water Resources Bulletin

CONTENTS

1. NITRATES

Abegglen, D.E., Wallace, A.T., and Williams, R.E. The Effect of Drain Wells on the Ground Water Quality of the Snake River Plain. Idaho Bur. Mines and Geol., Pamph. No. 148, 51 pp. (1970).

Disposal into aquifer, bacterial pollution problems, recommendations.

+ + +

Ackermann, W.C. Nitrate in Water Supplies. Illinois Water Survey, Technical Letter No. 6, pp. 1-3 (1966).

In Illinois 25% of all water samples from free wells of 50-feet depth or less contain an excessive concentration of nitrates.

+ + +

Anon. The Nitrate Dilemma. *Cross Section* 16, No. 6, pp. 1-4 (1970).

Ogallala summary.

+ + +

Baars, J.K. Experiences in the Netherlands with Contamination of Ground Water. (In *Ground Water Contamination, Proc. of the 1961 Symposium*), USDHEW, PHS, Robert A. Taft San. Eng. Cent., Tech. Report W61-5, pp. 56-63 (1961).

Case history of nitrate and coliform in sand.

1

Behnke, J.J., and Haskell, E.E., Jr. Ground Water Nitrates
Distribution beneath Fresno, California. *J. Amer. Water
Works Assoc.* 60, pp. 477-480 (1968).

+ + +

Cash, J.G. *et al.* Nitrates in Water Supplies, Field Crops
and Ruminant Nutrition. College of Agriculture, Coop.
Extension Service, Univ. of Illinois, Urbana, 13 pp.
(1967).

25% of all water samples from wells 50-feet deep or less
contain excessive concentrations of nitrates.

+ + +

Chalk, P.M., and Keeney, D.R. Nitrate and Ammonium Contents
of Wisconsin Limestone. *Nature* 229, No. 5279, p. 42 (1971).

+ + +

Chemerys, J.C. Effect of Urban Development on Quality of
Ground Water in Raleigh, North Carolina. USGS Professional
Paper 575-B, pp. 212-216 (1967).

Ten of sixty-two wells studied have excessive nitrate and
chloride.

+ + +

Crabtree, K.T. Nitrate Variation in Ground Water. Univ.
of Wisconsin (Madison), Water Resources Cent., Supplemental
Report, 60 pp. (1970).

Extensive nitrates in Manthon County (Wisconsin).

+ + +

Crabtree, K.T. Nitrate and Nitrite Variation in Ground
Water. Univ. of Wisconsin (Madison), Dept. of Natural
Resources, Tech. Bull. No. 58, 24 pp. (1972).

Ground water, nitrogen compounds, water wells, nitrogen
cycle, kinetics, nitrates, nitrites (Wisconsin).

Dawes, J.H., Larson, T.E., and Harmeson, R.H. Nitrate Pollution of Water. (In *Frontiers in Conservation, Proc. of the 24th Annual Meeting*), Soil Conservation Society of America, pp. 94-102 (1969).

+ + +

deLaguna, W. Chemical Quality of Water--Brookhaven National Laboratory and Vicinity, Suffolk County, New York. USGS Bull. 1156-D, 73 pp. (1964).

This report tabulates and interprets 300 chemical analyses of water samples from wells, lakes and rivers in the area. Higher concentration of radioactivity in surface waters has spread, evaluated as an effect of recent Nevada bomb tests. Widespread contamination of ground water was noted as a result of leaching of nitrates from fertilizers in farmed areas, and local contamination from cesspools was experienced.

+ + +

Feth, J.H. Nitrogen Compounds in Natural Water--A Review. *Water Resources Research* 2, No. 1, pp. 41-58 (1966).

Comprehensive review.

+ + +

Fitzsimmons, Lewis, Taylor, and Busch. Nitrogen, Phosphorus and Other Inorganic Materials in Waters in a Gravity-Irrigated Area. *Trans. Am. Soc. Agr. Eng.* 15, No. 2, pp. 292-295 (1972).

Agricultural runoff, nitrogen, phosphorus, irrigation, surface water, ground water, Boise Valley, Idaho.

+ + +

Gad, G., and Naumann, K. Harmful Contents of Nitrate in Water Supply. *Gas-u. Wasserfach*, pp. 684-685 (1956).

General discussion.

+ + +

Gale, H.S. Origin of Nitrates in Cliffs and Ledges. Mining and Scientific Press (San Francisco, Calif.), No. 115, pp. 676-678 (1917).

Examples from Iowa and Minnesota.

George, W.O., and Hastings, W.W. <u>Nitrate in the Ground Water of Texas.</u> *Trans. Am. Geophys. Union* <u>32</u>, No. 3, pp. 450-456 (1951a).

Most nitrate in wells less than 200-feet deep.

+ + +

Gillham, R.W., and Webber, L.R. <u>Nitrogen Contamination of Ground Water by Barnyard Leachates.</u> *J. Water Pollution Control Federation* <u>41</u>, pp. 1752-1762 (1969).

Concentration of nitrate related to ground water flow distribution.

+ + +

Glandon, L.R., Jr., and Beck, L.A. <u>Monitoring Nutrients and Pesticides in Subsurface Agriculture Drainage.</u> (In *Collected Papers* regarding nitrates in agricultural wastewater), U.S. Dept. Int., FWQA, Water Pollution Control Research Series 13030 ELY 12/69, pp. 53-79 (1969).

+ + +

Goldberg, M.C. <u>Sources of Nitrogen in Water Supplies.</u> In *Agricultural Practices and Water Supply,* Iowa State Univ. Press, Ames, Iowa, pp. 94-124 (1970).

+ + +

Green, L.A., and Walter, P.J. <u>Nitrate Pollution of Chalk Waters.</u> *J. Soc. Water Treat. Exam.* <u>19</u>, pp. 169-182 (1970).

High levels of nitrate found west of Eastbourne.

+ + +

Harmeson, R.H., and Larson, T.E. <u>Existing Levels of Nitrates in Waters--The Illinois Situation.</u> *Source and Control, Proc. of the 12th San. Eng. Conf.,* Eng. Pub. Office, University of Illinois, Urbana, pp. 27-39 (1970).

+ + +

Harmeson, R.H., Sollo, F.W., and Larson, T.E. <u>The Nitrate Situation in Illinois.</u> *J. Am. Water Works Assoc.* <u>63</u>, pp. 303-310 (1971).

Jacobson, R.L., and Langmiur, D. The Chemical History of
Some Spring Waters in Carbonate Rocks. *Ground Water* 8,
No. 3, pp. 5-9 (1970).

Most of the dissolved solids, including pollutants such
as chloride and nitrate, were added to spring waters
during ground water flow.

+ + +

Johnson, W.R., Ittihadich, F., Daum, R.M., and Pillsburg,
A.F. Nitrogen and Phosphorus in Tile Drainage Effluents.
Soil Sci. Soc. Am. Proc. 29, p. 287 (1965).

+ + +

Keeley, J.W., and Scalf, M.R. Aquifer Storage Determination
by Radiotracer Techniques. *Ground Water* 7, No. 1 (1969).

Comparison to DDT and nitrate.

+ + +

Keeney, D.R. Nitrates in Plants and Waters. *J. Milk and
Food Tech.* 33, No. 10, pp. 425-432 (1970).

+ + +

Keller, W.D., and Smith, G.E. Ground Water Contamination
by Dissolved Nitrate. *Geol. Soc. Am. Spec. Papers* 90,
59 pp. (1967).

+ + +

Kor, B.D., and Schneider, R.A. Delano Nitrate Investigation.
California Dept. Water Resources, Bull. No. 143-6, 42 pp.
(1968).

+ + +

Larson, T.E. Occurrence of Nitrate in Well Waters. Univ.
of Illinois, Urbana, Water Resources Center, Proj. 65-056,
15 pp. (1966).

23% of samples taken from shallow wells contained 45 mg/l
or more nitrate. Excessive concentrations are more likely
in drift wells than sandstone or limestone.

Lenain, A.F. The Impact of Nitrate on Water Use. *J. Am. Water Works Assoc.* 58, No.10, pp. 1049-1054 (1967).

+ + +

Longenecker, D.E. Far-West Texas Irrigation Waters Contain Nitrogen. *Texas Agr. Prog.* 9, No. 5, p. 12 (1963).

+ + +

McDermott, J.H., Kabler, P.W., and Wolf, H.W. Health Aspects of Toxic Material in Drinking Water. *Am. J. Public Health* 61, pp. 2269-2276 (1971).

Nitrate, zinc, mercury, arsenic and selenium.

+ + +

McGaughey, P.H. Nitrates in Water Supply--The Problem. *Source and Control, Proc. of the 12th San. Eng. Conf.,* Univ. of Illinois (Urbana), College of Engineering Publication, pp. 1-5 (1970).

+ + +

McMillian, L.G., and Hauser, V.L. Field Evaluation of Potential Pollution from Ground Water Recharge. *Water Well J.* 23, No. 8, pp. 23-24 (1969).

Ogallala nitrates.

+ + +

Meisler, Harold. Hydrogeology of the Carbonate Rocks of the Lebanon Valley, Pennsylvania. *Penn. Geol. Surv. Bull.* W-18 (4th Series), 81 pp. (1963).

Case history of nitrate pollution.

+ + +

Meisler, Harold. Hydrology of the Carbonate Rocks of the Lancaster 15-Minute Quadrangle, Pennsylvania. *Penn. Geol. Surv. Prog. Rep.* 171, 36 pp. (1966).

Case history of nitrate pollution.

Muschter, W.Z. <u>Nitrate Contents in Drinking Water and Two
Cases of Methamoglobinaimia in Berlin Caused by Well Water.</u>
Ges. Hyg. <u>8</u>, pp. 781-788 (1962); *Zembl. Bakt. Parasit. Kde.,
I., Ref.* <u>196</u>, p. 339 (1964).

+ + +

Murphy, L.S., and Gosch, J.W. <u>Nitrate Accumulation in
Kansas Ground Water.</u> Kansas Water Resources Res. Inst.
(Manhattan), Proj. Completion Rept., 56 pp. (1970).

+ + +

Nichols, M.S. <u>Nitrates in the Environment.</u> *J. Am. Water
Works Assoc.*, pp. 1319-1327 (1965).

A review of the literature.

+ + +

Nightingale, H.I. <u>Statistical Evaluation of Salinity and
Nitrate Content and Trends beneath Urban and Agricultural
Area--Fresno, California.</u> *Ground Water* <u>8</u>, No. 1, pp. 22-28
(1970).

Salinity of ground water beneath urban areas increases
with time; nitrate content increases with time in both
urban and agricultural areas.

+ + +

Nishimura, G.H. *et al.* <u>Nitrates in the South Central Coastal
Area of Southern California.</u> California Dept. Water
Resources, Memo Rept., 45 pp. (1971).

+ + +

Palmquist, W.N., and Hall, F.R. <u>Reconnaissance of Ground
Water Resources in the Blue Grass Region, Kentucky.</u>
USGS, WSP 1833, 39 pp. (1961).

Table 3 (Chemical Analyses of Ground Water) shows 11
wells with 50 ppm nitrate or more (maximum 587).

+ + +

Perlmutter, N.M., and Koch, E. <u>Preliminary Hydrogeologic
Appraisal of Nitrate in Ground Water and Streams, Southern
Nassau County, Long Island, New York.</u> USGS Professional
Paper 800-B, pp. 225-235 (1972).

Price, E.F. Genesis and Scope of Interagency Cooperative
  Studies of Control of Nitrates in Subsurface Agricultural
  Wastewaters.   U.S. Dept. Interior, FWQA, WPCRS, 13030
  ELY 12/69, pp. 1-14 (1969).

+ + +

Reginald, H., Young, F., Lav, S., and Burbank, N.C. Travel
  of ABS Ammonia Nitrogen with Percolating Water Through
  Saturated Oahu Soils.   Univ. of Hawaii, Water Resources
  Research Cent., Tech. Rept. 1, 54 pp. (1967).

  Laboratory study.

+ + +

Richter, J.C. High Nitrate Contents in Well Waters.
  Gesundheitswes. u. Desinfekt., 55, No. 5/6, p. 76 (1963);
  Zbl. Bakt., *I. Ref.*, 193, p. 84 (1964).

+ + +

Robeck, G.G. Microbial Problems in Ground Water. *Ground
  Water* 7, No. 3, pp. 33-35 (1969).

  Disease outbreaks; nitrate concentration buildup in semi-
  closed circuit.

+ + +

Scalf, M.R., Hauser, V.L., McMillion, L.C., *et al.* Fate of
  DDT and Nitrate in Ground Water. U.S. Dept. Agriculture
  (Ada, Oklahoma and Bushland, Texas), 46 pp. (1968).

+ + +

Scalf, M.R., Hauser, V.L., McMillion, L.C., Dunlap, W.J.,
  and Keeley, J.W. Fate of DDT and Nitrate in Ground Water.
  Public Works 100, No. 5, pp. 152-156 (1968).

  Recovery of nitrate, DDT, and tritium after being added
  to water used to recharge an aquifer in Texas.

+ + +

Scalf, M.R., Dunlap, W.J. *et al.* Movement of DDT and Nitrate
  During Ground Water Recharge. *Water Resources Research*
  5, No. 5, pp. 1041-1052 (1969).

  Ogallala aquifer; DDT absorbed into aquifer.

Schmidt, E.L. Soil Nitrification and Nitrates in Waters. *Public Health Rep.* 71, No. 5, pp. 497-503 (1956).

Examples of nitrate contamination from Minnesota.

+ + +

Schmidt, K.D. The Distribution of Nitrate in Ground Water in the Fresno-Clovis Metropolitan Area, San Joaquin Valley, California. Ph.D. Dissertation, Univ. of Arizona, 348 pp. (1971).

Main sources of nitrate in ground water are sewage waste effluent, system tanks effluent, and winery wastewaters.

+ + +

Schmidt, K.D. The Use of Chemical Hydrographs in Ground Water Quality Studies. Water Resources Assoc., Arizona Sect., Tucson, Arizona. *Hydrogeology and Water Resources in Arizona and Southwest* 1, pp. 211-223 (1971).

Changes with time of properties of water used to study nitrate in ground water of Fresno-Clovis metropolitan area, California.

+ + +

Schmidt, K.D. Nitrate in Ground Water of the Fresno-Clovis Metropolitan Area, California. NGWQS, Proc., USEPA, No. 16060 GRB 08/71, pp. 144-158 (1971).

High concentration due to industrial and municipal pollution.

+ + +

Schwille, F. High Nitrate Content in the Water of Wells in the Mosel Valley Between Trier and Koblenz. *Gas-u. Wasserfach* 110, pp. 35-44 (1969).

High nitrate content due to vineyards.

+ + +

Shearer, Goldsmith, Young, Kearns, and Tamplin. Methemoglobin Levels in Infants in an Area with High Nitrate Water Supply. *Am. J. Public Health* 62, No. 9, pp. 1174-1180 (1972).

Hematopoutic pathology, nitrogen compounds, water wells,
water sampling, water contamination, California data,
high nitrate water, methemoglobinemia infants.

+ + +

Stetson, C.L. <u>Nitrates in Ground Waters of the San Joaquin
Valley</u>. California Dept. Water Resources, Memo Rept.,
47 pp. (1970).

+ + +

Stetson, C.L. <u>Nitrates in Ground Waters of the Central
Coastal Area</u>. California Dept. Water Resources, Memo
Rept., 16 pp. (1971).

+ + +

Smith, G.E. <u>Nitrate Pollution of Water Supplies</u>. Univ.
of Missouri, *Proc., 3rd Annual Conf. on Trace Substances
in Environmental Health 1969,* pp. 273-287 (1970).

In Missouri, 42% of samples contaminated 5.00 or less
ppm nitrate; some shallow wells more than 300 ppm.

+ + +

Smith, W.D., Hill, R.O., and Walker, W.H. <u>Washington
County Nitrate Study - Phase 1</u>. Illinois Water Survey,
Open-File Rept., 15 pp. (1970).

+ + +

Stewart, B.A., Viets, F.G., Jr., Hutchinson, G.L., Kemper,
W.D., and Clark, F.E. <u>Distribution of Nitrates and Other
Water Pollutants Under Fields and Corrals in the Middle
South Platte Valley, Colorado</u>. U.S. Dept. Agriculture,
Res. Ser., ARS 41-134, 206 pp. (1967a).

+ + +

Stewart, B.A., Viets, F.G., Jr., Hutchinson, G.L., and
Kemper, W.D. <u>Nitrate and Other Water Pollutants Under
Fields and Feedlots</u>. *Env. Sci. Tech.* <u>1</u>, No. 9, pp. 736-
739 (1967b).

South Platte River and nitrates.

Tenorio, P., and Young, R.H.F. The Effect of Irrigation Return Water upon Ground Water Quality. *Proc., 5th Int. Conf. Water Pollution Research 1970* 2, Paper No. HA-20, 14 pp. (1971).

Hawaiian Islands; nitrates induced into basal aquifer.

+ + +

Thompson, G.L. Ground Water Resources of Nararro County, Texas. Texas Water Devel. Board, Rept. 160, 63 pp. (1972).

Of 96 samples, 34 contained more than 45 mg/l nitrate (of which 26 came from Midway group and Taylor). Max -- 2190 mg/l.

+ + +

Tikhe, D. Drinking Water and Nitrates Symposium. Visweswaraya Regional College of Engineering, Institution of Engineers, Nagpur,India (February,1972), 311 pp. (1972).

Potable water, nitrates, surface water, ground water contamination, sources, effects, removal.

+ + +

Turk, Heil, Kreitler, and Jones. Nitrate Contamination of Ground Water in Runnels County, Texas. *Proc., 5th Annual Conf.*, Univ. of Missouri (Columbia) Env. Health Cent., Extension Division (June-July, 1972), D.D. Hempshell, ed., pp. 153-163 (1972).

Nitrogen compounds, ground water, water contamination, public health, nitrates, Runnels County, Texas.

+ + +

Walker, E.H. Ground Water Resources of the Hopkinville Quadrangle, Kentucky. USGS WSP 1328, 98 pp. (1956).

Contamination widespread; 12% of ground water samples under nitrate analysis exceed recommended standard; 50% show coliform bacteria Mississippian limestone.

+ + +

Walker, W.H., Peck, T.R., and Lembke, W.D. Farm Ground Water Nitrate Pollution--A Case Study. ASCE Ann. & Nat. Env. Eng. Mtg. (October, 1972), Houston, Texas. Preprint 1842, Illinois Water Survey Reprint Series 228, 25 pp. (1972).

Walker, W.H. Illinois Ground Water Nitrate Pollution.
Illinois Water Survey, Open-File Rept. for presentation
at Illinois Pollution Control Board Hearing on Livestock
Waste Regulations.

+ + +

Ward, P.C. Existing Levels of Nitrate in Water--The
California Situation. *Proc., 12th San. Eng. Conf. on
Nitrate and Water Supply, Source and Control* (February,
1970), Univ. of Illinois (Urbana), College of Eng. Publ.,
pp. 14-26 (1970).

+ + +

Well, G.J., and Webber, L.R. Soil Characteristics and
Subsurface Sewage Disposal. *Can. J. Public Health* 61,
pp. 47-54 (1970).

Case histories of nitrate and phosphate at Coral Lake,
Ontario, cottage and uncontrolled sewage disposal.

+ + +

Wells, --. The Effect of Unlined Treated Sewage Storage
Ponds on Water Quality in the Ogallala Formation. U.S.
Nat. Tech. Inf. Service, Govt. Rept. Annou. 72(7), p. 139
(1972).

+ + +

Witzel, S., McCoy, E., Attoe, O.J., Polkowski, L.B., and
Crabtree, K.T. Nitrogen Cycle in Surface and Subsurface
Waters. Univ. of Wisconsin, Water Res. Cent., Tech. Rept.
(1968).

+ + +

Woodward, L. Ground Water Contamination in the Minneapolis
and St. Paul Suburbs. (In *Ground Water Contamination,
Proc. of the 1961 Symposium*), USDHEW, PHS, Robert A. Taft
San. Eng. Cent., Tech. Rept. W61-5, pp. 66-71 (1961).

Nitrate, ABS case history.

2. HEAVY METALS AND INDUSTRIAL WASTES

Andreyev, P.F., Bugrov, N.M., and Glebivskaya, V.S. Iso-topic Composition of Lead in Natural Waters. *Geochem. Int.* 4, No. 3, pp. 551-556 (1967).

Ground water of Kazakhstan and the rocks through which these waters percolate contain lead of similar composition.

+ + +

Bennett, R.R., and Meyer, R.R. Geology and Ground Water Resources of the Baltimore Area, Maryland. Maryland Dept. Geology, Mines, and Water Res., Bull. 4, 2573 pp. (1952).

Industrial wastes have caused local contamination with chromium and sulfuric acid.

+ + +

Bettolo, G.B.M., Piladino, S., Scarafoni, G.S., Villa, L. and Visintin, B. Research in Water Supply and Pollution by the Institute Superiore de Sanita. *Int. Conf. Water for Peace*, U.S. Govt. Printing Office, Washington, D.C., Vol. 3, 1967, pp. 714-723 (1968).

Case history of chromium in ground water near Milan.

+ + +

Burt, E.M. The Use, Abuse and Recovery of a Glacial Aquifer. NGWQS, Proc., USEPA, No. 16060 GRB 08/71, pp. 159-166 (1971).

Industrial pollution, decontamination procedures, Michigan.

13

Cheremisinoff, P.N., and Habib, Y.H. Cadmium, Chromium,
Lead, Mercury: A Plenary Account for Water Pollution
(Part 1 -- Occurrence of Toxicity and Detection). *Water
and Sewage Works* 119, No. 7, pp. 73-86 (1972).

+ + +

Csanady, M. Spread of Pollution by Heavy Metals and
Cyanides in Ground Water. *Magy. Tudom., Akad. Orv. Tudom,
Osztal. Kozl.,* 18, pp. 481-493 (1968); *Chem. Abstr.* 68,
p. 7848 (1968).

+ + +

Davids, H.W., and Lieber, W. Underground Water Contamination by Chromium Wastes. *Water and Sewage Works* 98,
No. 12 (1951).

Chromium contamination in Nassau County was first reported
in 1942. One family in Nassau County was drinking water
with 25 ppm hexavalent chromium as recently as 1951 without
any deleterious effects. The quality of ground water in
the contaminated area should continue to improve since
pollution has stopped.

+ + +

Deutsch, M. Incidents of Chromium Contamination of Ground
Water in Michigan. (In *Ground Water Contamination, Proc.
of the 1961 Symposium*), USDHEW, PHS, Robert A. Taft San.
Eng. Cent., Tech. Report W61-5, pp. 98-104 (1961).

+ + +

Dunlap, W.J., Crosby, R.L., McNabb, J.F., Bledsoe, B.E.,
and Scalf, M.R. Investigation Concerning Probable Impact
of NTA on Ground Water. USEPA, WPCRS 16060 GHR 11/71,
60 pp. (1971).

NTA survives infiltration, may transport heavy metals--
laboratory study.

+ + +

Effenberger, E. Pollution of Ground Water by Cyanide.
Arch. Hyg. Bakl. 148, pp. 271-287 (1964); *Bull. Hyg. Lond.*
39, p. 947 (1964).

Fink, B.E. Investigation of Ground Water Contamination
by Cotton Seed Delinting Acid Waste, Terry County, Texas.
Texas Water Comm. Rept. LD-0864, 24 pp. (1964).

+ + +

Flynn, J.M., Andrioli, A., and Guerrera, A.A. Study of
Synthetic Detergents in Ground Water. *J. Am. Water Works
Assoc.* 50, No. 12, pp. 1551-1562 (1958).

The water supplies of 186 homes were sampled by the
Suffolk County Department of Health. The study began in
1955 when complaints about a yellow color in the water
were received. Subsequently, analyses showed hexavalent
chromium from plating plant wastes in the water. At this
time complaints about frothy water were also made. These
conclusions were reached: when wells and septic tanks
are placed on small plots, synthetic detergents will
appear. With the continued addition of synthetic detergents to ground water reservoirs, the problem becomes
more acute.

+ + +

Friberg, L., Piscatov, M., and Nordberg, G. Cadmium in the
Environment. The Chemical Rubber Company, Cleveland,
Ohio, 166 pp. (1971).

Extensive bibliography.

+ + +

Golera, G.A., Polyakev, V.A., and Nechayeva, T.P. Distribution and Migration of Lead in Ground Water. *Geokhimiya,*
No. 3, pp. 344-357 (1970); *Geochem. Int.* 7, pp. 256-268
(1970).

+ + +

International Federation of Doctor-Engineers and Engineer
Doctors in Science. Proceedings, International Study
Conference on Water Problems. (May-June,1967) Paris,
France, 362 pp. (1967).

Full text, six sections. One section covers pollution
of ground and surface waters by chromium or other heavy
metals (treatment of wastewaters, pollution of ground
waters with chromium in Italy).

Koppe, P., and Gieber, G. The Danger of Public Water Supplies through the Pollution of Ground Water with Arsenic. *Stadlerydiene* 16, pp. 241-245 (1965).

Hydrogeological and chemical investigation of a town in Germany, tracing source of pollution of ground water by arsenic compounds from ammunition dumped after World War I.

+ + +

Leiber, M., and Welsch, F.W. Contamination of Ground Water by Cadmium. *J. Am. Water Works Assoc.* 46, No. 6, pp. 541-547 (1954).

Case history on Long Island.

+ + +

Leiber, M. Ground Water Pollution by Industrial Wastes, (Discussion of a paper by F.W. Welsch). *Sewage Ind. Wastes* 27m, No. 9, pp. 1069-1072 (1955).

+ + +

Leiber, M., Perlmutter, N.M., and Frauenthal, H.L. Cadmium and Hexavalent Chromium in Nassau County Ground Water. *J. Am. Water Works Assoc.* 56, No. 6, 9 pp. (1964).

A case history.

+ + +

Little, A.D., Inc. Water Quality Criteria Data Book . Vol. 2: Inorganic Chemical Pollution of Fresh Water. USEPA, WPCRS 1801 ODPV 07/71, U.S. Govt. Printing Office, 288 pp. (1971).

+ + +

McDermott, J.H., Kabler, P.W., and Wolf, H.W. Health Aspects of Toxic Material in Drinking Water. *Am. J. Public Health* 61, pp. 2269-2276 (1971).

Nitrate, zinc, mercury, arsenic and selenium.

+ + +

Michigan Department of Health. Unique Pollution of a Well by Chromium. *Mich. Water Works News* 21, No. 3 (1956).

Perlmutter, N.M., Lieber, M., and Frauenthal, H.L. Movement of Waterborne Cadmium and Hexavalent Chromium Wastes
in South Farmingdale, Nassau County, Long Island, New
York. USGS Professional Paper 475-C, Art. 105, 6 pp.
(1963).

Case history brought up to date.

+ + +

Perlmutter, N.M., and Lieber, M. Disposal of Plating
Wastes and Sewage Contaminants in Ground Water and Surface Water, South Farmingdale (Massapequa Area), Nassau
County, Long Island, New York. USGS WSP 1879-G, 74 pp.
(1970).

+ + +

Piper, A.M., Garrett, A.A. *et al.* Native and Contaminated
Ground Waters in the Long Beach - Santa Ana Area, California.
USGS WSP 1136, 320 pp. (1953).

Includes contamination by oil field and industrial waste.

+ + +

Schews, L.D. Blast Furnace Gas Sludge Containing Cyanide
as Cause of Ground Water Pollution. *Gas-u. Wasserfach*
110, pp. 702-706 (1969) (in German).

+ + +

Stetson, Carl L. Arsenic in Ground Waters of the San
Joaquin Valley. Calif. Dept. Water Resources, Memo Rept.,
11 pp. (1970).

+ + +

Welsh, W.F. Ground Water Pollution from Industrial Wastes.
Sewage Ind. Wastes 27, No. 9, pp. 1065-1069 (1955).

Ground water pollution from industrial wastes is creating
a dangerous situation in sections of Long Island. Chromium
and cadmium were both found in high concentrations in the
ground water of Nassau County, Long Island, New York.

+ + +

Wormold, B. Arsenic in Wells in Northeastern California.
Calif. Dept. Water Resources, Memo Rept., 19 pp. (1970).

3. PESTICIDES, HERBICIDES AND ORGANIC WASTES

Beran, F., and Guth, S.A. Organic Insecticides in Various Soils, with Particular Reference to Possible Ground Water Pollution. *Pflanzenschutzberichte* 33, pp. 65-117 (1965); *Chem. Abstr.* 65, p. 17635 (1966).

+ + +

Burtschell, R.H., Rosen, A.A., and Middleton, F.M. Two Cases of Organic Pollution of Ground Waters. (In *Ground Water Contamination, Proc. of the 1961 Symposium*), USDHEW PHS, Robert A. Taft San. Eng. Cent., Tech Rept. W61-5, pp. 115-117 (1961).

Locations not specified.

+ + +

Dugan, P.R., Pfister, R.M., and Sprague, M.L. Bibliography of Organic Pesticide Publications having Relevance to Public Health and Water Problems. New York Dept. Health, Research Rept., 10, Part 2, 122 pp. (1963).

+ + +

Dunlap, W.J., Crosby, R.L., McNabb, J.F. *et al.* Probable Impact of NTA on Ground Water. NGWQS, Proc., USEPA, No. 16060 GRB 08/71, pp. 201-211 (1971).

Substitute for phosphate, ground water contamination, experimental study.

Dunlap, W.J., Crosby, R.L., McNabb, J.F., Bledsoe, B.E., and Scalf, M.R. Investigation Concerning Probable Impact of NTA on Ground Water. USEPA, WPCRS 16060 GHR 11/71, 60 pp. (1971).

NTA survives infiltration, may transport heavy metals (laboratory study).

+ + +

Eye, J.D. Aqueous Transport of Dieldrin Residues in Soils. *Diss. Abstr.* 27, B, pp. 3548-3549 (1967).

Investigation of potential dieldrin pollution. Dieldrin was studied for solubility, absorption, and time required for transportation.

+ + +

Faust, H. Organic Compounds in Aquatic Environments. (New York: Marcel Dekker, Inc.), 659 pp. (1971).

Organic compounds, water resources, wastewater disposal, industrial wastes.

+ + +

Glandon, L.R., Jr., and Beck, L.A. Monitoring Nutrients and Pesticides in Subsurface Agriculture Drainage. (In *Collected Papers Regarding Nitrates in Agricultural Wastewater*), U.S. Dept. Interior, FWQA, WPCRS 13030 ELY 12/69, pp. 53-79 (1969).

+ + +

Grigaropoulas, S.G., and Smith, J.W.J. Trace Organics in Missouri Subsurface Waters. *J. Am. Water Works Assoc.* 60, pp. 586-596 (1968).

A study was made on the use of the carbon extraction procedure to recover organic contaminants from three sources of ground water (a spring and two wells).

+ + +

Grigaropoulas, S.G., Smith, J.W.J., and Mathews, J.R. Trace Organic Subsurface in Missouri Waters. *Proc., Trace Subst. Env. Health*, No. 3, pp. 251-271 (1970).

Grigaropoulas, S.G., and Smith, J.W. Trace Organics in
Subsurface Water. In *Organic Compounds in Aquatic
Environments* (New York: Marcel Dekker, Inc.), pp. 95-
118 (1971).

Contamination, pollution, geochemistry, experimental
studies.

+ + +

Ishizaki, K., Burbank, N.C., Jr., and Law, S. Effects of
Soluble Organics on Flow Through Thin Cracks of Basaltic
Lava. Univ. of Hawaii, Water Res. Cent., Tech. Rept. 16,
56 pp. (1967).

Organics tend to plug thin cracks.

+ + +

Jettmar, H.M. Pollution of Ground Water by Substances
Difficult to Decompose. *Ost. Wasserw.* 9, p. 56 (1957).

BHC applied to surface soil contaminated the shallow
well (33 ft) water supply of Basel, Switzerland, and
produced a disagreeable taste. Also cited are two cases
of ground water pollution by trichloroethylene. In one
case contamination of a well by seepage from a waste pit
at a distance of 150 meters persisted for at least four
years.

+ + +

Jordan, D.G. Ground Water Contamination in Indiana. *J.
Am. Water Works Assoc.* 54, pp. 1213-1220 (1962).

Case histories of ground water contamination by bacteria,
saline water, sulfate, chromate (four cases, one in limestone), and phenol.

+ + +

Keeley, J.W., and Scalf, M.R. Aquifer Storage Determination
by Radiotracer Techniques. *Ground Water 7*, No. 1 (1969).

Compared to DDT and nitrate.

Lewallen, M.J. Pesticide Contamination of a Shallow Bored Well in the Southeastern Coastal Plain. *Ground Water* 9, p. 6 (1971).

DDT and toxaphene persisted for four years.

+ + +

McFarland, P.H., and Krone, R.B. Report of Investigation of Travel of Pollution. Calif. State Water Pollution Board, Publ. No. 11, 218 pp. (1954).

+ + +

Middleton, M., and Walton, G. Organic Chemical Contamination of Ground Water. (In *Ground Water Contamination, Proc. of the 1961 Symposium*), USDHEW, PHS, Robert A. Taft San. Eng. Cent., Tech. Rept. W61-5, pp. 50-56 (1961).

Review article; cites case histories in Ohio, West Virginia, Illinois, Texas and Colorado.

+ + +

Scalf, M.R., Hauser, V.L., McMillion, L.C., *et al.* Fate of DDT and Nitrate in Ground Water. U.S. Dept. Agriculture, Ada, Oklahoma, and Bushland, Texas, 46 pp. (1968).

+ + +

Scalf, M.R., Hauser, V.L., McMillion, L.C., Dunlap, W.J., and Keeley, J.W. Fate of DDT and Nitrate in Ground Water. *Public Works* 100, No. 5, pp. 152-156 (1968).

Recovery of nitrate, DDT and tritium after being added to water used to recharge an aquifer in Texas.

+ + +

Scalf, M.R., Dunlap, W.J., *et al.* Movement of DDT and Nitrate During Ground Water Recharge. *Water Resources Research* 5, No. 5, pp. 1041-1052 (1969).

Ogallala aquifer, DDT absorbed into aquifer.

Schnieder, A.D., Weiss, A.F., and Jones, O.R. Movement and Recovery of Herbicides in the Ogallala Aquifers. Int. Center for Arid and Semi-Arid Land Studies, Special Report No. 39, pp. 219-226 (1970).

+ + +

Semenov, A.D., Nemtseva, L.I., Brazhnikova, L.V., Demchenko, A.S., *et al.* Composition of Organic Substances in Flood Waters Forming in Small Catchment of the Arid Zones. *Soviet Hydrology*, Selected Paper No. 5, pp. 554-557 (1967).

+ + +

Smith, J.W., and Gregaropoulas, S.G.T. Toxic Effects of Odorous Trace Organics. *J. Am. Water Works Assoc.* 60, pp. 969-979 (1968).

Organic micro-pollutants recovered from surface and ground water, a spring and two wells were studied. Their acute and long-term toxic effects were evaluated.

+ + +

Strohl, G.W. A Case of Far-Reaching Ground Water Pollution Caused by Pesticides and Detergents. *Gesundheitsingenieur* 87, pp. 108-114 (1966).

Case history (Transvaal, South Africa).

+ + +

Tsapko, V.V., and Kupyrov, V.N. Underground Water Contamination with Insecticides. *Gig. San.* 33, No. 8, pp. 6-9 (1968).

Laboratory study on benzene hexachloride, hepachloride, and mononitrotoluene (English summary).

+ + +

Walker, T.R. Ground Water Contamination in the Rocky Mountain Arsenal Area, Denver, Colorado. GSA Bulletin 72, p. 489 (1961).

The contamination originated from a waste basin that held chlorate from 1943 to 1950 and 2,4-D type compounds from 1950 to 1957. The chlorate pollution has migrated about 5 miles and the 2,4-D type about 3.5 miles.

Yakubova, R.A., and Bashirov, R.R. Chemization of Agriculture and the Problem of Water Hygiene Fields. *Gig. San.* <u>31</u>, No. 9, pp. 21-23 (1966).

Various poisonous chemicals widely used in agriculture (DDT, aldrin, intrathion, and methylmercaptopos) were detected in amounts exceeding the maximum permissible level in surface and ground water.

4. URBANIZATION

Anon. <u>Hydrology in the Urban Environment</u>. Am. Geol. Inst.
(Washington, D.C.), AGI Short Course Lecture Notes,
37 pp. (1970).

Effects of urbanization on hydrological system, increased
yield, peak discharge, reduce lag time and recharge.

+ + +

Bailey, G.D. <u>The Fragipan Menace and Comprehensive Planning</u>
<u>in Megalopolis</u>. *Proc., Assoc. Am. Geographers* 3, pp. 25-28
(1971).

+ + +

Boughton, W.C. <u>Effects of Land Management on Quality and</u>
<u>Quantity of Available Water</u>. Australian Water Resources
Council, Research Report No. 120, 330 pp. (1970).

+ + +

Brendel, K., Haendel, D., Hohl, R., *et al*. <u>Some Geologic</u>
<u>Problems in the Relations between Environment and Man in</u>
<u>the Densely Populated Industrial Area of Holle-Leipzig</u>.
Geol. Berl. 21, No. 4-5, pp. 608-622 (1972).

Economics, subsidence due to leaching, ground water
pollution in Germany.

Chemerys, J.C. Effect of Urban Development on Quality of
Ground Water, Raleigh, North Carolina. USGS Professional
Paper 575-B, pp. 212-216 (1967).

Ten of sixty-two wells studied have excessive nitrate and
chloride.

+ + +

Detwyler, T.R., Marcus, G., *et al.* Urbanization and
Environment. Duxburg Press (Belmont, California), 287 pp.
(1972).

+ + +

Dial, D.C. Public Water Supplies in Louisiana. Louisiana
Dept. Public Works, Bosie Records Rept. 3, 460 pp. (1970).

Use of ground water by 246 municipalities; chemical
analyses of water samples.

+ + +

Dion, N.P. Some Effects of Land Use Changes on the Shallow
Ground Water System in the Boise-Nampa Area, Idaho.
Idaho Dept. Reclam., Water Information Bull., No. 26,
p. 45 (1972).

Data 1953-1970. Increased urbanization has generally
had little effect on ground water.

+ + +

Gabrysch, R.K. Development of Ground Water in the Houston
District, Texas, 1966-1969. Texas Water Development Board,
Rept. No. 152, 24 pp. (1972).

+ + +

Leopold, L.B. Hydrology for Urbanland Planning. In *Man
and his Physical Environment*. G.D. McKenzie and R.O.
Utgurd, eds. (Minneapolis, Minn: Burgess Publishing
Co., 1972), pp. 43-45.

+ + +

Nightingale, H.I. Statistical Evaluation of Salinity and
Nitrate Content and Trends beneath Urban and Agricultural
Area -- Fresno, California. *Ground Water* 8, No. 1,
pp. 22-28 (1970).

Salinity of ground water beneath urban areas increases with time; nitrate increases with time in both urban and agricultural areas.

+ + +

Nishimura, G.H., *et al.* Nitrates in the South Central Coastal Area of Southern California. California Dept. Water Resources, Memo Rept., 45 pp. (1971).

+ + +

Remson, I. Hydrology and Disposal Problems in Urban Areas. In *Environmental Planning and Geology*, USGS Office Res. Technol., Washington, D.C., pp. 36-41 (1971).

Contamination in watersheds, water supplies, Brandywine Creek, Chester County, Pa.

+ + +

Schaake, J.C., Jr. Water and the City. In *Urbanization and Environment: The Physical Geography of the City.* T. Ditwyler and M. Marcus, eds. (North Scituate, Mass: Duxburg Press, 1972), pp. 97-133.

Ground water, drainage patterns, uses and demands of water in cities.

+ + +

Scrudato, R.J. Surface Water Effects on Municipal Water Supplies in Logan County, West Virginia. *Geol. Soc. Am., Abs. with Prog.* 5, No. 5, p. 433 (1973).

Water wells along major streams and tributaries contain high sulfate, manganese, iron, and coliform bacteria.

+ + +

Sommers, D.A. Put Hydrology into Planning. In *Man and his Physical Environment.* G.D. McKenzie and R.O. Utgurd, eds. (Minneapolis, Minn: Burgess Publishing Co., 1972), pp. 279-284.

Spieker, A.M. Urbanization and the Water Balance. *Proc.
of Symposium on Water Balance in North America,* June 23-26,
1969, Banff, Alberta, Canada. Am. Wat. Res. Assoc.,
Urbana, Illinois, pp. 182-187 (1969).

+ + +

Thomas, H.E., and Schneider, W.J. Water as an Urban
Resource and Nuisance. USGS Circ. 601-D, 9 pp. (1970).

5. PETROLEUM AND PETROLEUM PRODUCTS

Cederstrom, D.J. The Arlington Gasoline Contamination
Problem. U.S. Dupl. Rept., 5 pp; Abstr. USGS WSP 1492,
p. 28 (1947).

+ + +

Crain, L.J. Ground Water Pollution from Natural Gas and
Oil Production in New York. N.Y. Water Resources Comm.,
Rept. Inv. R15, 15 pp. (1969).

Waste disposal on ground leakage and spillage from wells,
abandoned fields.

+ + +

Hoffman, B. Dispersion of Soluble Hydrocarbons in Ground
Water Stream. *Proc. 5th Int. Conf. Water Pollution
Research 1970*, $\underline{2}$, Paper No. HA 76, 8 pp. (1971).

+ + +

Hopkins, H.T. The Effect of Oilfield Brines on the Potable
Ground Water in the Upper Big Pitman Creek Basin, Kentucky.
Kentucky Geol. Surv., Rept. Inv., Series 10, No. 4, 36 pp;
Publ. Health Eng. Abstr. 1964, $\underline{44}$, p. 235 (1963).

+ + +

Kolle, W., and Sontheimer, H. Pollution of Underground
Water by Mineral Oil Products. *Brennst. -Chem.*, $\underline{50}$,
pp. 123-129 (1969).

29

Case histories of oil spills.

+ + +

Krieger, H. Experimental Investigation into Pollution of
Ground Water by Liquid Fuels. *Dt. Gewasserk. Mitt.*,
Special Issue No. 45 (1963).

+ + +

Krieger, R.A., and Henrickson, G.E. Effects of Greensburg
Oilfield Brines on the Streams, Wells and Springs of the
Upper Green River Basin, Kentucky. Kentucky Geol. Surv.,
Rept. Inv., Series 10, No. 2, 36 pp. (1960).

+ + +

Krieger, R.A. Ground Water Contamination in the Greensburg
Oilfield, Kentucky. (In *Ground Water Contamination, Proc.
of the 1961 Symposium*), USDHEW, PHS, Robert A. Taft San.
Eng. Cent., Tech. Rept. W61-5, pp. 91-97 (1961).

Detailed case history.

+ + +

Matis, J.R. Petroleum Contamination of Ground Water in
Maryland. NGWQS, Proc., USEPA, No. 16060 GRB 08/71,
pp. 57-61 (1971).

Procedures for investigating; remedies.

+ + +

McKee, J.E., Laverty, F.B., and Hertel, R.M. Gasoline in
Ground Water. *J. Water Pollution Control Federation* 44,
pp. 293-302 (1972).

+ + +

Meinhard, H. The Protection of Water from Risks Occurring
through the Storage of Liquids Hazardous to Water.
Wasserwertschaft, Stuttg. 55, pp. 33-35 (1965).

Effects of oil on ground waters.

Miller, L.M. Contamination by Processed Petroleum Products.
(In *Ground Water Contamination, Proc. of the 1961 Symposium*),
USDHEW, PHS, Robert A. Taft San. Eng. Cent., Tech. Rept.
W61-5, pp. 117-119 (1961).

+ + +

Mull, R. The Migration of Oil Products in the Subsoil
with Regard to Ground Water Pollution by Oil. *Proc. 5th
Int. Conf. Water Pollution Research 1970*, 2, Paper No.
HA-7a, 8 pp. (1971).

+ + +

Powell, W.J., Carroon, L.E., and Avrett, J.R. Water
Problems Associated with Oil Production in Alabama.
Alabama Geol. Surv., Circ. 22, 63 pp. (1964).   .

+ + +

Prier, H. Effects of Seepage of Mineral Oil Products on
the Ground Water Stratum. *Gesundheitsingenieur* 88,
pp. 145-149 (1967).

+ + +

Sackmann, L.A., and Zilliox, L. Pollution and Protection
of Underground Water. *Techgs. Sci. Munic.* 58, pp. 223-235
(1963).

A method enabling easy recognition of ground water polluted by petrol.

+ + +

Sackmann, L.A. Pollution of Ground Water by Hydrocarbons.
Eau 52, pp. 303-305 (1965).

+ + +

Zillrox, M. Etude sur Modèles Physiques du Mécanisme de
la Pollution des Eaux Souterrains par Liquides Miscibles
(Saumures) et Non Miscibles (Hydrocarbures). Société
Hydrotechnique de France, Comité Technique No. 93 (Proc.,
June 17-18, 1971), Houille Blanche 26(8), pp. 723-730
(1971).

Ground water, brine wastes, petroleum, crude oil, physical model, brine seepage.

+ + +

Zimmerman, W. The Effect of Mineral Oil on Ground Water. *Gewasserschutz Wass. Abwass.*, No. 3, pp. 23-41 (1970).

Gives details as the effect on plants, soil, and bacteria. (In German)

6. MICROORGANISMS

Abegglen, D.E., Wallace, A.T., and Williams, R.E. The
Effect of Drain Wells on the Ground Water Quality of the
Snake River Plain. Idaho Bur. Mines and Geol., Pamph.
148, 51 pp (1970).

Disposal into aquifer; bacterial pollution problems;
recommendations.

+ + +

Baars, J.K. Experiences in the Netherlands with Contami-
nation of Ground Water. (In *Ground Water Contamination,
Proc. of the 1961 Symposium*), USDHEW, PHS, Robert A. Taft
San. Eng. Cent., Tech. Rept. W61-5, pp. 56-63 (1961).

Case history of nitrate and coliform in sand.

+ + +

Bogan, R.H. Problems Arising from Ground Water Contamination by Sewage Lagoons at Tieton, Washington. (In *Ground
Water Contamination, Proc. of the 1961 Symposium*), USDHEW,
PHS, Robert A. Taft San. Eng. Cent., Tech. Rept. W61-5,
pp. 83-87 (1961).

Coliform in basalt.

+ + +

Butler, R.G., Orlob, G.T., and McGaughey, P.H. Underground
Movement of Bacterial and Chemical Pollutants. *J. Am.
Water Works Assoc.* **46**, p. 97 (1954).

Drewry, W.A., and Eliassen, R.S. Virus Movement in Ground
Water. *J. Water Pollution Control Federation* 40, pp. 257-
271 (1968).

Experiments with models on the movement of viruses in
ground water (bacteriophages).

+ + +

Ehrlich, G.G., Ehlke, T.A., and Vecchioli, J. Microbio-
logical Aspects of Ground Water Recharge-Injection of
Purified Chlorinated Sewage Effluent. USGS Professional
Paper 880-B, pp. 241-245 (1972).

+ + +

Esfandiari, F. Procedure and Results of Ground Water
Pollution Studies in Teheran Area, India. Congress of
Tokyo (Asian Regional Congress), *Int. Assoc. Hydro-
geologists Memoires* 9, pp. 3-7 (1972).

210 wells, chemical and bacterial analyses, good quality,
light pollution.

+ + +

Farkasdi, G., Golwer, A., Knoll, A., Mattess, K.H., and
Schneider, W. Investigations on the Microbiology and
Hygiene of Ground Water Pollution Downstream from Waste
Deposits. *Stadtehygiene* 20, pp. 25-31 (1969).

High counts on soil samples downstream at depths of
3-5 meters. Water same depth, counts not low. (In
German).

+ + +

Heukelekian, H., and Voelker, R.A. Factors Influencing
the Bacteriological Behavior of Private Home Wells.
Int. J. Air and Water Pollution 7, Nos. 4-5, pp. 303-
316 (1963).

+ + +

Jordan, D.G. Ground Water Contamination in Indiana. *J.
Am. Water Works Assoc.* 54, pp. 1213-1220 (1962).

Case histories of ground water contamination by bacteria, saline water, sulfate, chromate (four cases, one in limestone), and phenol.

+ + +

Knorr, M. The Hygienic Aspects of Toxic Resistant Substances in Soil and Ground Water. *Gesundheitsingenieur* 87, pp. 326-336 (1966).

+ + +

Kudryavtseva, B.M. Survival and Spread of *B. Coli* Group in Ground Waters. *Gig. San.* 53, No. 6, pp. 14-19 (1970).

Coliform survive 3-3.5 months. (In Russian, English summary).

+ + +

Luthy, R.G. New Concepts for Iron Bacteria in Water Wells. *Water Well J.* 1, No. 3, pp. 29-30 (1964).

+ + +

Mallmann, W.L., and Mack, W.N. Biological Contamination of Ground Water. (In *Ground Water Contamination, Proc. of the 1961 Symposium*), USDHEW, PHS, Robert A. Taft San. Eng. Cent., Tech. Rept. W61-5, pp. 35-43 (1961).

Review article.

+ + +

Nehrhorn, A. Nutrient-Deficient Substrate Used in Bacterial Counts of Ground Water. *Gesundheitsingerieur* 88, (1968).

20 counts/ml = slightly polluted. (In German).

+ + +

Randell, A.D. Movement of Bacteria from a River to a Municipal Well--A Case History. *J. Am. Water Works Assoc.* 62, No. 11, pp. 716-720 (1970).

Reeves, R.D., Rawson, J., and Blakey, J.F. Chemical and
Bacteriological Quality of Water at Selected Sites in
the San Antonio Area, Texas (August 1968 - April 1972).
Edwards Under Ground Water District, San Antonio, Texas,
63 pp. (1972).

+ + +

Robeck, G.G. Microbial Problems in Ground Water. *Ground
Water* 7, No. 3, pp. 33-35 (1969).

   Disease outbreaks, nitrate concentration buildup in
semi-closed circuit.

+ + +

Scrudato, R.J. Surface Water Effects on Municipal Water
Supplies of Logan County, West Virginia. *Geol. Soc. Am.,
Abs. with Prog.* 5, No. 5, p. 433 (1973).

Water wells along major streams and tributaries contain
high sulfate, manganese, iron and coliform bacteria.

+ + +

Slavnia, G.P. Naphthalene Oxidizing Bacteria in Ground
Waters from Oil Storage Deposits. *Mikrobiologiya* 34,
pp. 128-132 (1965).

+ + +

Stiles, C.W., and Crohurst, H.R. Principles Underlying
the Movement of *B. Coli* in Ground Water with the Resultant
Pollution on Wells. *Public Health Rept.* 38, p. 1350
(1923).

+ + +

Thomas, H.E. Artificial Recharge of Ground Water by the
City of Boutiful, Utah. *Trans. Am. Geophys. Union* 30,
No. 4, pp. 529-542 (1949a).

Because bacteria find optimum living conditions at the
water table, wells furnishing domestic supplies should
be tightly cased from the surface down to several feet
below the water table. Article cites bacterial contamination in deep water wells at Houston, Texas, and includes
selected references on bacterial pollution of ground water.

Vecchioli, J., Ehrlich, G.G., and Ehlke, T.A. Travel of Pollution-Indicator Bacteria through the Magothy Aquifer, Long Island, New York. USGS Professional Paper No. 800-B, pp. 237-239 (1972).

+ + +

Vogt, J.E. Infectious Hepatitis Outbreak in Posen, Michigan. (In *Ground Water Contamination, Proc. of the 1961 Symposium*), USDHEW, PHS, Robert A. Taft San. Eng. Cent., Tech. Rept. W61-5, pp. 87-91 (1961).

+ + +

Walker, E.H. Ground Water Resources of the Hopkinville Quadrangle, Kentucky. USGS WSP 1328, 98 pp. (1956).

Contamination widespread; 12% of ground water samples under nitrate analysis exceed recommended standard; 50% show coliform bacteria Mississippian limestone.

+ + +

Weber, G. Chemical and Bacteriological Investigation of Ground Water During an Epidemic of Hepatitis. *Osterr. Wasser (Austria)* 10, p. 110 (1958); *Public Health Eng. Abstr.* 39, No. 11, p. 27 (1959).

7. LIQUID WASTE AND SEWAGE DISPOSAL

American Water Works Association, Task Group 245-OR.
Underground Waste Disposal and Control. *J. Am. Water
Works Assoc.* 49, pp. 1334-1342 (1957).

+ + +

Andres, B.D. Effect of Synthetic Detergents on the Ground
Waters on Long Island, New York. N.Y. Water Pollution
Control Board, Research Rept. No. 6, 18 pp. (1960).

Ground water is polluted by laundry and laundramat waste
at Mastic, Farmingville, Deer Park, and other areas. The
pollution increases the total dissolved solids content
by 50 to 150 ppm. The apparent willingness to abandon
the shallow aquifer to pollution is questioned. Figures
are presented showing the plan views at sites investigated,
geologic sections, and depth to which observation wells
were constructed. A table of the laundry survey presents
basic data.

+ + +

Bair, D.C., and Wesner, G.M. Reclaimed Waste for Ground
Water Recharge. *Water Resources Bull.* 7, pp. 991-1001
(1971).

Reclaimed water acceptable for domestic use after travel
through 500 feet of a confined aquifer.

Boen, D.F., Bunts, J.H., Jr., and Currie, R.J. Study of
Reutilization of Wastewater Recycle through Ground Water.
USEPA, WPCRS No. 16060 DDZ, 176 pp. (1971).

Feasibility study--Hemet-San Jacinto Valley, California.

+ + +

Bogan, R.H. Problems Arising from Ground Water Contamination by Sewage Lagoons at Tieton, Washington. (In *Ground
Water Contamination, Proc. of the 1961 Symposium*), USDHEW,
PHS, Robert A. Taft San. Eng. Cent., Tech. Rept. W61-5,
pp. 83-87 (1961).

Coliform in basalt.

+ + +

Born, S.M., and Stephenson, D.A. Hydrological Considerations in Liquid Waste Disposal. *Soil and Water Conservation* 24, No. 2, pp. 52-55 (1969).

+ + +

Bouwer, H., Lance, J.C., and Rice, R.C. Renovating Sewage
Effluent by Ground Water Recharge. (In *Hydrology and
Water Resources in Arizona and the Southwest* 1, pp. 225-
244), Am. Water Resources Assoc., Arizona Sect., Tuscon
(1971).

Description of experiments made at Flushing Meadows
project area, Phoenix, Arizona.

+ + +

Caldwell, E.L. Pollution Flow from Pit Latrines Where an
Impervious Statum Closely Underlies the Flow. *J. Infectious
Diseases* 61, p. 270 (1937).

+ + +

Caldwell, E.L., and Parr, L.W. Ground Water Pollution and
the Borehole Latrine. *J. Infectious Diseases* 61, p. 148
(1937).

+ + +

Caldwell, E.L. Pollution Flow from a Pit Latrine Where
Permeable Soils of Considerable Depth Exist Below the Pit.
J. Infectious Diseases 62, p. 225 (1938).

Campenni, L.G. Synthetic Detergents in Ground Waters, Part 1. *Water and Sewage Works*, May 1961, 4 pp. (1961).

Briefly reviews the nature of the contamination and specifically the situation on Long Island.

+ + +

Chuck, R.T., and Lum, D. Disposal of Waste Effluent in Coastal Limestone Aquifers in a Tropical Island Environment. *Proc. 5th Int. Conf. Water Pollution Research, 1970*, 2, Paper No. HA-19, 8 pp. (1971).

Describes proposed use of three deep wells at Waimanalo, Oahu, Hawaii.

+ + +

Cohen, P., Vaupel, D.E., and McClymonds, N.E. Detergents in the Streamflow of Suffolk County, Long Island, New York. USGS Professional Paper 750-C, pp. 210-214 (1971).

+ + +

Cotteral, J.A., and Norris, D.P. Septic Tank Systems. *Am. Soc. Civil Engineering Proc.*, 95, Paper 6735; *J. San. Engineering Div.*, No. SA 4, pp. 715-746 (1969).

Recovery, history and theory of septic tanks; recommendations.

+ + +

Flynn, J.M. Impact of Suburban Growth on Ground Water in Suffolk County, New York. (In *Ground Water Contamination, Proc. of the 1961 Symposium*), USDHEW, PHS, Robert A. Taft San. Eng. Cent., Tech. Rept. W61-5, pp. 71-82 (1961).

This report discusses the pollution of the ground water in the glacial deposits of Suffolk County with emphasis on detergent contaminants.

+ + +

Flynn, J.M. Water Resources--Growing Pains. *Water and Sewage Works* 112, No. 5 (1965).

Wastewater contamination of ground water supplies in Suffolk County, New York, cited as impetus for implementing sewer systems and abandoning ground water disposal systems.

ABS major contaminant. Drought defined as low supply
of good quality water.

+ + +

Freethey, G.W., and Waltz, J.P. Hydrogeologic Evaluation
of Pollution Potential at Mountain Dwelling Sites. *Geol.*
Soc. Am., Abs. with Prog., Part 5, Rocky Mountain Section,
p. 25 (1969).

+ + +

Gomez-Pallete y Rivas, F. Disposal of Liquid Wastes by
Means of Injection into the Earth. Spain Inst. Geol.
Min., *Bol. Geol. Min.* 82, No. 5, pp. 59-67 (1971).

Basic scheme for liquid waste disposal. (English
summary).

+ + +

Goolsby, D.A. Hydrogeochemical Effects of Injecting Wastes
into Limestone Aquifers near Pensacola, Florida. *Ground*
Water 9, No. 1, pp. 13-19 (19 71).

+ + +

Hajeh, B.F. Chemical Interactions of Wastewater-Soil
Environment. *J. Water Pollution Control Federation* 41,
No. 10, pp. 1175-1186 (1969).

Soil with wastewater relationships.

+ + +

Hansen, G.M. Sewage Disposal and its Effects on Ground
Water beneath Livermore Valley. *Eng. Geol.* 1, No. 2,
pp. 22-36 (1964).

+ + +

Harmeson, R.H. ABS in Ground Water. (In *Ground Water*
Contamination, Proc. of the 1961 Symposium), USDHEW, PHS,
Robert A. Taft San. Eng. Cent., Tech. Rept. W61-5, pp.
190-193 (1961).

Examples from Illinois.

Henry, H.R. The Effects of Temperature and Density Gradient upon the Movement of Contaminants in Saturated Aquifers. *Proc., Symposium on Geochemistry, Precipitation, Evaporation, Soil Moisture, Hydrometeorology* (Bern, Switzerland), International Association of Scientific Hydrologists, Pub. No. 78, pp. 54-65 (1968).

Experimental studies, simplified geometric model; thermal diffusivity 50 times larger than usual because of dispersion.

+ + +

Hoffman, J.A., and Spiegal, S.J. Chloride Concentration and Temperature of Water from Wells in Suffolk County, Long Island, N.Y., 1928-1953. New York Water Power and Control Commission, Bull. GW-38, 55 pp. (1958).

Fertilizer and sewage contribute chloride to ground water.

+ + +

Keeley, J.W. Problems of Ground Water Pollution by Brine. *Public Works* 97, No. 3, pp. 149-162 (1966).

+ + +

Korotchansky, A.N., and Mitchell, J. Waste Disposal in Deep Aquifers. *Int. Geol. Cong., Proc.,* Section 11, No. 24, pp. 282-296 (1972).

Application of recent techniques to the industrial waste in the Paris Basin, France.

+ + +

Kostin, P.P. The Sanitary Protection Zone in the Underground Storage of Liquid Industrial Wastes. *Z. Angew. Geol.* 18, No. 4, pp. 171-173 (1972).

+ + +

Krone, R.B., McGaughey, P.H., and Gotaas, H.B. Direct Recharge of Ground Water with Sewage Effluents. *Proc., ASCE,* Paper No. 1335, 83 pp. (1957).

Kumagas, J.S.   Infiltration and Percolation of Sulfides
and Sewage Carbonaceous Matter. University of Hawaii,
Water Resources Research Center, Tech. Rept. 7, 58 pp.
(1967).

+ + +

Lauman, H.E.   Detergent Cocktails. *Water Well J.* 16,
No. 6 (1962).

Describes extent of syndet pollution in Suffolk County,
Long Island, New York.

+ + +

Lawton, G.W.J.   Detergents in Wisconsin Waters. *J. Am.
Water Works Assoc.* 59, pp. 1327-1334 (1967).

+ + +

Leiber, M.   Ground Water Pollution by Industrial Wastes.
(Discussion of a paper by F.W. Welsch), *Sewage Ind.
Wastes* 27m, No. 9, pp. 1069-1072 (1955).

+ + +

Lloyd, J.W., Drennon, D.S.H., and Bennell, B.M.U.   A Ground
Water Recharge Study in Northeastern Jordan. *Proc. Inst.
Civil Engineering* 35, pp. 615-631 (1966).

+ + +

Lohnert, E.   The Pollution of Ground Water in the Elke
Valley (Free Town and Merchant Town Hamburg). *Gas-u.
Wasserfach* 10, pp. 1171-1177 (1969).

Waste disposal case history (in German).

+ + +

Lutzen, E.E.   Ground Water Contamination from Urbanization
in a Carbonate Terrane. Assoc. Eng. Geol., *Ann. Mett.,
Prog. Abstr.*, No. 15, p. 30 (1972).

Spring pollution, sewage, correction -- Missouri.

Marsh, J.H. Design of Disposal Wells. In *Man and His Physical Environment,* G.D. McKenzie and R.O. Utgurd, eds. (Minneapolis, Minn: Burgess Publishing Co.), pp. 159-164 (1972).

+ + +

McGaughey, P.H., and Orlob, G.T. Fate of Synthetic Detergents in Ground Waters. *Proc., 3rd Annual Sanitary and Water Resources Engineering Conference* (Vanderbilt Univ., Nashville, Tenn.), P.A. Krenkel, ed., pp. 88-96 (1965).

+ + +

Meinhard, H. The Protection of Water from Risks Occurring through the Storage of Liquids Hazardous to Water. *Wasserwertschaft, Stuttg.* 55, pp. 33-35 (1965).

Effects of oil on ground waters.

+ + +

Moyer, R., and Rognon, Ph. Disposal of Industrial Wastewaters in Deep Geologic Layers. French Bur. Geo. Min., Bull., Series 2, Section 3, No. 2, pp. 3-6 (1971).

Methods, examples, Paris region.

+ + +

Murashita, T. Problems in Artificial Recharge through Artesian Wells in Industrial Areas. Congress of Tokyo (Asian Regional Congress), *Int. Assoc. Hydrogeologists Memoires* 9, pp. 99-100 (1971).

Experimental studies in Japan.

+ + +

Muzzi, A., Borgioli, P., and Lardati, A. Synthetic Detergents as an Index of Pollution of Ground Water. *Nuovi Annali Iq. Microbiol.* 19, pp. 494-508 (1969).

150 samples, 10 wells, 1 year, 15 meters deep.

Newwell, I., and Almquist, F. Contamination of Ground
Water by Synthetic Detergents. *J. NEWWA* 74, p. 61
(1960).

+ + +

Perlmutter, N.M., Lieber, M., and Frauenthal, H.L. Con-
tamination of Ground Water by Detergents in a Suburban
Environment, South Farmingdale, Long Island, New York.
USGS Professional Paper 501-C, pp. 170-175 (1964).

+ + +

Perlmutter, N.M., and Guerrera, A.A. Detergents and
Associated Contaminants in Ground Water at Three Public
Supply Well Fields in Southwestern Suffolk County, Long
Island, New York. U.S. Geol. Surv., Water Supply Paper
2001-B (1970).

+ + +

Perlmutter, N.M., and Lieber, M. Disposal of Plating
Wastes and Sewage Contaminants in Ground Water and Sur-
face Water, South Farmingdale (Massapequa Area), Nassau
County, Long Island, New York. U.S. Geol. Surv., Water
Supply Paper 1879-G, 74 pp. (1970).

+ + +

Perlmutter, N.M., and Koch, E. Preliminary Findings of
the Detergent and Phosphate Contents of Water of Southern
Nassau County, Long Island, New York. USGS Professional
Paper 750-D, pp. 171-177 (1971).

+ + +

Peterson, F.L., and Hargis, D.R. Effect of Storm Runoff
Disposal and Other Artificial Recharge to Hawaiian Ghyben-
Herzberg Aquifers. University of Hawaii, Water Resources
Center, Tech. Rept. No. 54, 51 pp. (1971).

+ + +

Petri, L.R. The Movement of Saline Ground Water in the
Vicinity of Derby, Colorado. (In *Ground Water Contamination, Proc. of the 1961 Symposium*), USDHEW, PHS, Robert
A. Taft San. Eng. Cent., Tech. Rept. W61-5, pp. 119-121
(1961).

Contamination from disposal ponds.

+ + +

Piper, A.M. Disposal of Liquid Wastes by Injection Under-
    ground--Neither Myth nor Millenium. USGS Circ. 631,
    15 pp. (1969).

    General discussion.

+ + +

Piper, A.M. Disposal of Liquid Wastes by Injection Under-
    ground. In *Man and His Physical Environment*, G.D.
    McKenzie and R.O. Utgurd, eds., (Minneapolis, Minn:
    Burgess Publishing Co.), pp. 148-158 (1972).

+ + +

Preul, H.C. Contaminants in Ground Water near Waste
    Stabilization Ponds. *J. Water Pollution Control Federation*
    40, No. 4, pp. 659-669 (1968).

    Field study (three years), no problem in sandy soils.

+ + +

Reeder, L.R. Underground Waste Disposal Systems. *Geol.
    Soc. Am., Abs. with Prog.* 4, No. 4, p. 290 (1972).

+ + +

Reichenbaugh, R.C. Water Quality Aspects of Spraying
    Sewage on a Pasture near Lakeland, Florida. *Proc., Lat.
    Geol. Cong., (Hydrol.)*, Section 11, No. 24, p. 305 (1972).

+ + +

Ricciardi, G., and Tridenk, M. Pollution of Ground Water
    in Puglia Research for Anionic Detergents in the Waters
    of Some Wells Situated in the Areas of Bari. *Ig. Mod.*
    61, pp. 968-987 (1968).

    44% wells near Bari discharged syndets.

Sayre, A.N., and Stringfield, V.T. Artificial Recharge
of Ground Water Reservoirs. *J. Am. Water Works Assoc.*
40, p. 1152 (1948).

Phenols detected in ground water were traced to the
waste discharged from a plant producing a weed killer
containing mainly trichlorophenol. The phenols were
detected within 17 days of the time that the plant
began discharging its chlorinated effluent to the Rio
Hondo. The waste percolated down through the Rio Hondo
river bed to the ground water basin. This short-time
discharge of waste caused taste and odor problems that
were evident 4-1/2 to 5 years later.

+ + +

Scgicht, R.J. Feasibility of Recharging Treated Sewage
Effluent into a Deep Sandstone Aquifer. NGWQS, Proc.,
USEPA, No. 16060 GRB 08/71, pp. 20-35 (1971).

Ground water pollution in the Chicago area.

+ + +

Smith, H.F. Subsurface Storage and Disposal in Illinois.
NGWQS, Proc., USEPA, No. 16060 GRB 08/71, pp. 20-28 (1971).

Liquid and gas storage; fluid industrial waste disposal.

+ + +

Stundl, K. The Influence of Contaminated Surface Waters
on the Underground Waters. Int. Conf. Water Resources,
Washington, 1967, 3, pp. 897-905 (1968).

+ + +

Sturtevant, R.M., and Brimley, E.V. A Study of Ground
Water Supplies in Two Massachusetts Communities as
Related to Synthetic Detergents. *Sanitalk* 11, No. 1,
pp. 2-8 (1963).

+ + +

Suess, M.J. Retardation of ABS in Different Aquifers.
J. Am. Water Works Assoc. 56, No. 1, pp. 89-91 (1964).

Svore, J.H. Sewage Lagoons and Man's Environment. *Civil Engineering* 34, No. 9, pp. 54-56 (1964).

+ + +

Unklesbay, A.G., and Cooper, H.H., Jr. Artificial Recharge of Artesian Limestone at Orlando, Florida. *Econ. Geol.* 41, No. 4, Part 1, pp. 293-307 (1946).

Wells 200 feet or more in depth used to drain streets, control lake levels, and dispose of sewage liquid wastes.

+ + +

Vogt, J.E. Infectious Hepatitis Outbreak in Posen, Michigan. (In *Ground Water Contamination, Proc. of the 1961 Symposium*), USDHEW, PHS, Robert A. Taft San. Eng. Cent., Tech. Rept. W61-5, pp. 87-91 (1961).

Glacial drift.

+ + +

Walker, T.R. Ground Water Contamination in the Rocky Mountain Arsenal Area, Denver, Colorado. GSA Bull. No. 72, p. 489 (1961).

The contamination originated from a waste basin that held chlorate from 1943 to 1950 and 2,4-D type compounds from 1950 to 1957. The chlorate pollution has migrated about 5 miles and the 2,4-D type about 3.5 miles.

+ + +

Waller, R. Anionic Detergents in Ground Water Supplies. *Water Works News of New York State* (New York Dept. Health), 15, No. 4 (1960).

Shallow and deep wells show evidence of contamination.

+ + +

Walton, G. Effects of Pollutants in Water Supplies--ABS Contamination. *J. Am. Water Works Assoc.* 52, pp. 1354-1362 (1960).

Walton, G. Report on Analyses of Water Samples from Rocky
    Mountain Arsenal Area, Denver, Colorado. USDHEW, PHS,
    Robert A. Taft San. Eng. Center (1960).

+ + +

Walton, G. Public Health Aspects of the Contamination of
    Ground Water in the Vicinity of Derby, Colorado. (In
    Ground Water Contamination, Proc. of the 1961 Symposium),
    USDHEW, PHS, Robert A. Taft San. Eng. Cent., Tech. Rept.
    W61-5, pp. 121-125 (1961).

    Contamination from disposal ponds.

+ + +

Waltz, J.P. Methods of Geologic Evaluation of Pollution
    Potential at Mountain Homesites. NGWQS, Proc., USEPA,
    No. 16060 GRB 08/71, pp. 137-143 (1971).

    Septic tank system not suited for the terrain (Rocky
    Mountains, Colorado).

+ + +

Warman, J.C. Ground Water Waste Disposal Recharge and
    Reuse. NGWQS, Proc., USEPA, No. 16060 GRB 08/71, pp.
    36-44 (1971).

+ + +

Wayman, Cooper, Page, and Robertson. Behavior of Sur-
    factants and Other Detergent Components in Water and
    Soil-Water Environment. Federal Housing Administration,
    Tech. Studies Publication No. FHA 532, 136 pp. (1965).

+ + +

Weber, G. Chemical and Bacteriological Investigation of
    Ground Water During an Epidemic of Hepatitis. *Osterr.
    Wasser (Austria)* 10, p. 110 (1958); *Public Health Eng.
    Abstr.* 39, No. 11, p. 27 (1959).

+ + +

Well, G.J., and Webber, L.R. Soil Characteristic and
    Subsurface Sewage Disposal. *Canada J. Public Health* 61,
    pp. 47-54 (1970).

Case histories of nitrate and phosphate at Coral Lake, Ontario, cottage and uncontrolled sewage disposal.

+ + +

Wells, --. The Effect of Unlined Treated Sewage Storage Ponds on Water Quality in the Ogallala Formation. U.S. Nat. Tech. Inf. Service, Govt. Rept. Annou. 72(7), p. 139 (1972).

Nitrates, sewage, ground water contamination, storage lagoons, Ogallala Formation.

+ + +

Welsh, W.R. Ground Water Pollution from Industrial Wastes. *Sewage and Industrial Wastes* 27, No. 9, pp. 1065-1069 (1955).

Ground water pollution from industrial wastes to the ground water is creating a dangerous situation in sections of Long Island. Chromium and cadmium were both found in high concentrations in the ground water of Nassau County.

+ + +

Wilson, L.G. Investigations of the Subsurface Disposal of Waste Effluent at Inland Sites. U.S. Office Saline Water Research and Development, Progress Rept. 650, 106 pp. (1971).

Brackish waste disposal at inland sites.

8. LIMESTONE TERRANE

AIHS-UNESCO. <u>Hydrology of Fractured Rocks</u>. *Proc.,*
Dubrovnik Symposium, 2 volumes, 689 pp. (1965).

This collection of papers deals primarily with hydrology
of karst terranes and includes coverage of the following
categories:

1. Theory of the flow in karst
2. Hydraulic and hydrologic characteristics of the
    karst hydrological balance--geological character-
    istics
3. Karstic hydrologic regions
4. Salt encroachment in karstic regions
5. Natural and artificial recharge in karstic
    regions
6. Karstic hydrology in volcanic terrane
7. Exploitation
8. Influence on runoff
9. Karstic lakes
10. Geochemistry and erosion

+ + +

Anon. <u>James River-Wilson Creek Study, Springfield, Missouri</u>.
FWPCA, <u>1</u>, Robert A. Kerr Water Research Center, Ada,
Oklahoma, 60 pp. (1969).

Good case history of contamination in limestone terrane.

Bailey, B.L., and Malatino, A.M. Contamination of Ground
Water in a Limestone Aquifer in the Stevenson Area,
Alabama. Alabama Geological Survey, Circ. No. 76, 15 pp.
(1971).

+ + +

Baker, H., Jr. Florida Regulations Pertaining to Ground
Water Contamination. (In *Ground Water Contamination,
Proc. of the 1961 Symposium*), USDHEW, PHS, Robert A. Taft
San. Eng. Cent., Tech. Rept. W61-5, pp. 141-149 (1961).

Sites examples from 16 locations of ground water contamination.

+ + +

Bennet, G.D., and Giust, E.V. Ground Water in the
Tortuquero Area, Puerto Rico, As Related to Proposed
Harbor Construction. Puerto Rico Dept. Public Works,
Water Resource Bull. No. 10, 25 pp. (1972).

Agmamon limestone, water-table aquifer; harbor would
not appreciably alter amount of flow.

+ + +

Bowman, I., and Reeds, C.A. Water Resources of the East
St. Louis District. Illinois Geological Survey, Bull.
No. 5, 128 pp. (1907).

Describes contamination of karst water.

+ + +

Burdon, D.J., and Safadi, C.J. The Karst Ground Water of
Syria. *Hydrology* 2, pp. 324-347 (1964).

+ + +

Chalk, P.M., and Keeney, D.R. Nitrate and Ammonium Contents of Wisconsin Limestone. *Nature* 229, No. 5279,
p. 42 (1971).

+ + +

Chuck, R.T., and Lum, D. Disposal of Waste Effluent in
Coastal Limestone Aquifers in a Tropical Island Environment. *Proc., 5th Int. Conf. Water Pollution Research (1970)*
2, Paper No. HA-19, 8 pp. (1971).

Describes proposed use of three deep wells at Waimanalo, Oahu, Hawaii.

+ + +

Dreyfuss, M. Special Pollution Problems in Limestone Country. *Eau*, No. 53, pp. 79-82 (1966).

+ + +

Feder, G.L. Geochemical Survey of Trace Elements in Waters of Missouri. *Geol. Soc. Am., Abs. with Prog.* 5, No. 4, (North Central Section), p. 314 (1973).

+ + +

Foose, R.M. Ground Water Behavior in the Hershey Valley, Pennsylvania. *Geol. Soc. Am. Bull.* 64, No. 6, pp. 623-646 (1953).

Beekmantown limestone dewatering caused development of sink holes after mine began pumping 6500 gpm.

+ + +

Goolsby, D.A. Hydrogeochemical Effects of Injecting Wastes into Limestone Aquifers near Pensacola, Florida. *Ground Water* 9, No. 1, pp. 13-19 (1971).

+ + +

Gray, D.A. Ground Water Conditions of the Chalk of the Grimsby Area, Lincolnshire. U.S. Geological Survey, Water Supply Paper, Research Report No. 1 (1964).

+ + +

Green, L.A., and Walter, P.J. Nitrate Pollution of Chalk Waters. *J. Soc. Water Treatment Exam.* 19, pp. 169-182 (1970).

High levels of nitrate found west of Eastborne.

+ + +

Harvey, E.J., and Skelton, J. Hydrologic Study of a Waste Disposal Problem in a Karst Area at Springfield, Missouri. USGS Professional Paper 600-C, pp. 217-220 (1968).

Langmiur, D., Parizck, R.R., and Apgar, M.A. The Chemical
Interaction of Sanitary Landfill Leachates with Unsaturated
Soil in a Carbonate Rock Terrain. Pennsylvania Dept.
Health, Progress Report, 27 pp. (1970).

+ + +

Lutzen, E.E. Ground Water Contamination from Urbanization
in a Carbonate Terrane. Assoc. Eng. Geol., *Ann. Mett.,
Prog. Abstr.,* No. 15, p. 30 (1972).

Spring pollution, sewage, correction -- Missouri.

+ + +

Meisler, H. Hydrogeology of the Carbonate Rocks of the
Lebanon Valley, Pennsylvania. Pennsylvania Geological
Survey, Bull. No. W-18, 4th Series, 81 pp. (1963).

Case history of nitrate pollution.

+ + +

Meisler, H. Hydrology of the Carbonate Rocks of the
Lancaster 15-Minute Quadrangle. Pennsylvania Geological
Survey, Progress Report 171, 36 pp. (1966).

Case history of nitrate pollution.

+ + +

Ward, P.E., and Wilmoth, B.M. Ground Water Hydrology of
the Monogahela River Basin in West Virginia. West
Virginia Geol. and Econ. Survey, River Basin Bull. 1,
54 pp. (1968).

Aquifers in sandstone and limestone; shallow aquifer
contamination.

9. ANIMAL WASTES, FERTILIZERS, AND AGRICULTURAL PRACTICES

Anon. Fertilizers and Feedlots--What Role in Ground Water Pollution. *Agr. Research* 18, No. 6, pp. 14-15 (1969).

+ + +

Bourodimos, E.L., and Michna, L. Potential Contribution of Fertilizers to Ground Water Pollution. *Trans. Am. Geophys. Union* 53, No. 4, p. 368 (1972).

+ + +

Edwards, W.M., and Harold, L. Agricultural Pollution of Water Bodies. *Ohio J. Science* 70, No. 1, pp. 59-56 (1970).

+ + +

Fitzsimmons, Lewis, Taylor, and Busch. Nitrogen, Phosphorus, and Other Inorganic Materials in Waters in a Gravity-Irrigated Area. *Trans. Am. Soc. Agr. Eng.* 15(2), pp. 292-295 (1972).

Agricultural runoff, nitrogen, phosphorus, irrigation, Boise Valley (Idaho), surface water, ground water.

+ + +

Gillham, R.W., and Webber, L.R. Nitrogen Contamination of Ground Water by Barnyard Leachates. *J. Water Pollution Control Federation* 41, pp. 1752-1762 (1969).

Concentration of nitrate related to ground water flow
distribution.

+ + +

Glandon, L.R., Jr., and Beck, L.A. Monitoring Nutrients
and Pesticides in Subsurface Agricultural Drainage.
(In *Collected Papers* regarding nitrates in agricultural
wastewater), U.S. Dept. Int., FWQA, Water Pollution Control Research Series 13030 ELY 12/69, pp. 53-79 (1969).

+ + +

Gordon, G.V. Chemical Effects of Irrigation Return Water,
North Platte River, Western Nebraska. USGS Professional
Paper 550-C, pp. 224-250 (1966).

+ + +

Hoffman, J.A., and Spiegal, S.J. Chloride Concentration
and Temperature of Water from Wells in Suffolk County,
Long Island, New York, 1928-1953. New York Water Power
and Control Commission, Bull. GW-38, 55 pp. (1958).

Fertilizer and sewage contribute chloride to ground
water.

+ + +

Johnson, W.R., Ittihadich, F., Daum, R.M., and Pillsburg,
A.F. Nitrogen and Phosphorus in Tile Drainage Effluents.
Soil Sci. Soc. Am. Proc. **29**, p. 287 (1965).

+ + +

LeGrand, H.E. Movement of Agricultural Pollutants with
Ground Water. In *Agricultural Practices and Water
Quality* (Ames, Iowa: Iowa State University Press),
pp. 303-313 (1970).

+ + +

Mielke, L.N., Ellis, J.R., Swanson, N.P., *et al.* Ground
Water Quality and Fluctuation in a Shallow Unconfined
Aquifer Under a Level Feedlot. In *Relation of Agricul-
ture to Soil and Water Pollution* (Ithaca, New York:
Cornell University Press), pp. 34-40 (1970).

Mink, J.F. Excessive Irrigation and the Soils and Ground
Water of Oahu, Hawaii. *Science* 135, pp. 672-673 (1962).

+ + +

Nightingale, H.I. Statistical Evaluation of Salinity and
Nitrate Content and Trends Beneath Urban and Agricultural
Area--Fresno, California. *Ground Water* 8, No. 1, pp. 22-
28 (1970).

Salinity of ground water beneath urban area increases
with time; nitrate increases with time in both urban and
agricultural areas.

+ + +

Nishimura, G.H., *et al*. Nitrates in the South Central
Coastal Area of Southern California. California Dept.
Water Resources, Memo Rept., 45 pp. (1971).

+ + +

Price, E.F. Genesis and Scope of Interagency Cooperative
Studies of Control of Nitrates in Subsurface Agricultural
Wastewaters. (In *Collected Papers* regarding nitrates in
agricultural wastewater), U.S. Dept. Int., FWQA, WPCRS,
13030 ELY 12/69, pp. 1-14 (1969).

+ + +

Robertson, J.S., Croll, J.M., James, A., and Gay, J.
Pollution of Underground Water from Pea Silage. *Mon.*
Bull. Minst. Health 25, pp. 172-179 (1966).

+ + +

Schmidt, K.D. The Distribution of Nitrate in Ground Water
in the Fresno-Clovis Metropolitan Area, San Joaquin Valley.
California. Ph.D. Dissertation, University of Arizona,
348 pp. (1971).

Main sources of nitrate in ground water are sewage waste
effluent, system tank effluent, and winery wastewaters.

+ + +

Smith, G.E. Contribution of Fertilizers to Water Pollution.
2nd Compendium of Animal Waste Management, Paper No. 7,
16 pp. (1969).

Smith, H.F., Harmeson, R.H., and Larson, T.E. The Effect
of Commercial Fertilizer on the Quality of Ground Water.
Illinois Water Survey (paper presented at the *Symposium
on Pollution of Ground Water*), IASH, The 15th General
Assembly of the IUGG (August), Moscow, USSR (1971).

+ + +

Stewart, B.A., Viets, F.G., Jr., Hutchinson, G.L., Kemper,
W.D., and Clark, F.E. Distribution of Nitrates and Other
Water Pollutants Under Fields and Corrals in the Middle
South Platte Valley, Colorado. U.S. Dept. Agriculture,
Research Series ARS 41-134, 206 pp. (1967a).

+ + +

Stewart, B.A., Viets, F.G., Jr., Hutchinson, G.L., and
Kemper, W.D. Nitrate and Other Water Pollutants Under
Fields and Feedlots. *Env. Sci. Tech.* 1, No. 9, pp. 736-
739 (1967b).

South Platte River and nitrates.

+ + +

Tenorio, P., and Young, R.H.F. The Effect of Irrigation
Return Water upon Ground Water Quality. *Proc. 5th Int.
Conf. Water Pollution Research 1970*, 2, Paper No. HA-20,
14 pp. (1971).

Hawaiian Islands; nitrates induced into basal aquifer.

+ + +

Texas Tech University. Infiltration Rates and Ground Water
Quality Beneath Cattle Feedlots, Texas High Plain. U.S.
Water Quality Office, WPCRS 16060 EGS 01/71 GPO, 64 pp.
(1971).

+ + +

Walton, G. Public Health Aspects of the Contamination of
Ground Water in the South Platte River Basin in the
Vicinity of Henderson, Colorado (August 1959). USDHEW,
PHS, Robert A. Taft San. Eng. Cent. (1959).

Yakubova, R.A., and Bashirov, R.R. Chemization of Agriculture and the Problem of Water Hygiene Fields. *Gig. San.* <u>31</u>, No. 9, pp. 21-23 (1966).

Various poisonous chemicals widely used in agriculture (DDT, aldrin, intrathion, and methylmercaptopos) were detected in amounts exceeding the maximum permissible level in surface and ground water.

10. GEOLOGIC, HYDROLOGIC AND LEGAL ASPECTS

Albinent, M. Determination of the Protection Parameters
of Water Catchment Areas Destined for Public Consumption.
French Bur. Réch. Geol. Minières Bull., No. 4, Series 2,
Section 3, pp. 33-37 (1971).

+ + +

American Water Works Association, Task Group E 4-C. Control of Ground Water Disposal, A Progress Report. J. Am.
Water Works Assoc. 44, pp. 685-689 (1952).

+ + +

American Water Works Association, Task Group E 4-C. Findings
and Recommendations on Underground Wastes Disposal. J.
Am. Water Works Assoc. 45, pp. 1295-1297 (1953).

+ + +

Anon. Pollution of Ground Water (In Legal Control of Water
Pollution). Davis Law Review, University of California,
pp. 141-165 (1969).

+ + +

Bredehoett, J.D., and Pinder, G.F. The Application of the
Transport Equations to a Ground Water System. Hydrogeology
(Section 11), Int. Geol. Congress, Proc., No. 24, pp. 225-
263 (1972).

Waste disposal systems, simulation of field conditions, contamination, application to Brunswick, Georgia.

+ + +

Brown, R.H. Hydrologic Factors Pertinent to Ground Water Contamination. (In *Ground Water Contamination, Proc. of the 1961 Symposium*), USDHEW, PHS, Robert A. Taft San. Eng. Cent., Tech. Rept. W61-5, pp. 7-16 (1961).

General mathematical treatment.

+ + +

Brown, R.H. Hydrologic Factors Pertinent to Ground Water Contamination. *Ground Water* 2, No. 1, pp. 5-12 (1964).

+ + +

Castany, G. Importance of Geomorphologic Factors in Surface Water - Ground Water Interactions and Evaluation of Water Resources. *Hydrogeology (Section 11), Int. Geol. Congress Proc.*, No. 24, pp. 121-129 (1972).

Stratigraphic structure; pollution examples from France, North Africa, Venezuela and Spain.

+ + +

Crosby, J.W., III, Johnstone, D.L., Drake, C.H., and Fenton, R.L. Migration of Pollutants in a Glacial Outwash Environment. *Water Resources Research* 4, No. 5, pp. 1095-1114 (1968).

+ + +

Crosby, J.W., III, Johnstone, D.L., and Fenton, R.L. Migration of Pollutants in a Glacial Outwash Environment. *Water Resources Research* 7, No. 1, pp. 204-208 (1971).

+ + +

DeBuchananne, G.D., and LaMoreaux, P. Geologic Controls Related to Ground Water Contamination. (In *Ground Water Contamination, Proc. of the 1961 Symposium*), USDHEW, PHS Robert A. Taft San. Eng. Cent., Tech. Rept. W61-5, pp. 3-7 (1961).

Definitions of geologic terms.

Deutsch, M. <u>Ground Water Contamination and Legal Controls in Michigan.</u> USGS Water Supply Paper 1461 (Open-File Report, 1960), 79 pp. (1961).

+ + +

Deutsch, M. <u>Hydrogeologic Aspects of Ground Water Pollution.</u> *Water Well Journal* <u>15</u>, No. 9, pp. 10-11, 35-39 (1961).

Describes examples from Michigan.

+ + +

Fournelle, H.J., Day, E.K., and Page, W.B. <u>Experimental Ground Water Pollution at Anchorage, Alaska.</u> U.S. Public Health Service, *Public Health Rept.* <u>72</u>, 203 pp. (1957).

+ + +

Georgescu, D., and Fazekas, I. <u>Research on the Chemical Pollution of Ground Water.</u> Studi Aliment Apa. Inst., *Studii Cerc. Hidroteh* <u>2</u>, pp. 3-31 (1965).

Laboratory tests (English summary).

+ + +

Crabovnikov, V.A. <u>Application of Balance Calculation for Studying Dispersion Haloes of Matter in Ground Water.</u> *Sov. Geol.*, No. 2, pp. 134-144 (1972).

+ + +

Kaufman, W.J. <u>Inorganic Chemical Contamination of Ground Water.</u> (In *Ground Water Contamination, Proc. of the 1961 Symposium*), USDHEW, PHS, Robert A. Taft San. Eng. Cent., Tech. Rept. W61-5, pp. 43-50 (1961).

Review article.

+ + +

LeGrand, H.E. <u>Management Aspects of Ground Water Contamination.</u> *J. Water Pollution Control Federation* <u>36</u>, p. 1133 (1964b).

LeGrand, H.E. Hydrogeologic Factors Controlling Pollutant Movement in Shallow Ground. *Geol. Soc. Am., Abs. with Prog.* <u>4</u>, No. 2, pp. 86-87 (1972).

+ + +

Margat, J., and Monition, L. Quality of Water, Pollution, Prevention, Protection, Exposition, Remedies, Generalities. In *Qualité des eaux, Pollutions, première partie,* French Bur. Réch. Géol. Minières, Bull., Series 1, Section 3, No. 4, pp. 5-12 (1970).

+ + +

Mattess, G. Hydrologic Criteria for the Self-Purification of Polluted Ground Water. *Hydrogeology (Section 11), Int. Geol. Congress Proc.,* No. 24, pp. 296-304 (1972).

Chemical, physical and microbiological criteria, permeability, ground water velocity.

+ + +

McGaughey, P.H., and Krone, R.B. Report of Investigation of Travel of Pollution. California State Water Pollution Board, Publ. 11, 218 pp. (1954).

+ + +

McMillian, L.G. Aquifer Protection and Rehabilitation. NGWQS, Proc., Bull Session 4, USEPA, No. 16060 GRB 08/71, pp. 167-181 (1971).

+ + +

Miakin, Y.L. Investigation and Estimation of Resources Relative to Ground Water Protection. *Izd. Nedra (Moscow),* 109 pp. (1972).

Includes "pollution factors."

+ + +

Michels, Nabert, Udluft, and Zimmerman. Gutachten zur Frage Schutzes des Grundwasser gegan Verunreinigung durch Lagerflussigkeiten. (Expert Opinion on Questions of the Protection by Aquifers Against Contamination of Ground Water). Bundesministerium fur Atomkernenergie and Wasserwirtschaft, Bad Godesberg (1959).

Miller, R.F. <u>Differences in Soil Chemistry Induced by</u>
<u>Evaporation and Flow of Ground Water</u>. USGS Professional
Paper 650-D, pp. 225-259 (1969).

+ + +

Nelson, R.W. <u>Evaluating the Environmental Consequences</u>
<u>of Ground Water Contamination through the Location and</u>
<u>Arrival Time Distributions</u>. (Abstract) *EGS, Am. Geophys.*
Union <u>53</u>, No. 4, p. 377 (1972).

+ + +

Patze, D. <u>The Effect on Ground Water, Particularly Bank</u>
<u>Filtered, of Radioactivity from Surface Waters</u>. *Gewasserik.*
Mitt. <u>7</u>, pp. 44-46 (1963).

+ + +

Ruedisili, L.G. <u>Ground Water in Wisconsin--Quality Pro-</u>
<u>tection: Legal Controls and Management</u>. University of
Wisconsin (Madison), Water Resources Center, Hydraulic
and Sanitation Lab. (1972).

+ + +

Roberts, F.W. <u>Contamination of Rivers and Other Water</u>
<u>Supply Sources</u>. *Munie, Engineering Lond.* <u>143</u>, pp. 678-
685 (1966).

Some of the causes of pollution of rivers and ground
water.

+ + +

Romm, Ye.S. <u>Filtration Properties of Jointed Rocks</u>.
Izd-vo. (Moscow), 283 pp. (1966).

+ + +

Sackmann, L.A. <u>Protection of Ground Water Aquifers</u>. Con-
<u>trol of the Water Table of Aquifer of the Rhine Valley</u>
<u>and Antipollution Methods</u>. *Houille Blanche* <u>26</u>, No. 8,
pp. 717-729 (1971).

Rhine graben, quality, contamination sources, counter
measures (with English summary).

Sharp, R.E. <u>Liability of Landowners for Pollution of Percolating Water</u>. *Marq. Law Review* <u>39</u>, No. 2, pp. 119-134 (1955).

+ + +

Snel, M.J. <u>Protection of Ground Water</u>. *Eau*, No. 229, pp. 23-29 (1971).

Aquifer contamination, causes, remedies, Belgium examples, water-bearing sands (Tertiary/Pleist).

+ + +

Tyagi, A.K., and Todd, D.K. <u>Dispersion of Pollutants in Saturated Porous Media</u>. *EGS, Am. Geophys. Union* <u>52</u>, No. 11, p. 833 (1971).

+ + +

Warner, D.L. <u>Preliminary Field Studies Using Earth Resistivity Measurements for Delineating Zones of Contaminated Ground Water</u>. *Ground Water* <u>7</u>, No. 1, pp. 9-16 (1969).

Long Island, Western Texas; electrical resistivity as a method for evaluating variations in ground water quality.

+ + +

Webster, D.S., Proctor, J.F., and Marine, I.W. <u>Two-Well Tracer Test in Fractured Crystalline Rock</u>. USGS Water Supply Paper 1544-I, pp. 1-22 (1970).

Experiments with well injection.

+ + +

Weiss, C.M., and Okum, D.A. <u>Water Quality Technology-- Present Capabilities and Future Prospects</u>. *Int. Conf. Water for Peace*, U.S. Govt. Printing Office, Washington, D.C., Vol. 4, 1967, pp. 195-211 (1968).

11. GENERAL ARTICLES INCLUDING CASE HISTORIES AND POLLUTION FROM UNSPECIFIED SOURCES

Agnew, A. Chemical Contamination of Ground Water.
NGWQS, Proc. (Bull Session 2), USEPA No. 16060 GRB
08/71, pp. 62-75 (1971).

+ + +

Bowman, I. Problems of Water Contamination. (In *Underground Water Papers*, M.L. Fuller, ed.), USGS Water Supply
Paper 160, pp. 92-96 (1906).

+ + +

Brunotte, Gendrin, Hugon, and Simler. A Study of Ground
Water Pollution by Salt. *Proc., 5th Int. Conf. Water
Pollution Research 1970*, $\underline{1}$, Paper No. I-34, 13 pp.
(1971).

+ + +

Buydens, R. Zones of Protection for Ground Water Supplies.
Techq. Eau Assain. $\underline{16}$, No. 182, pp. 17-23 (1962).

Review of ground water pollution, Belgium.

+ + +

Elrick, D.E., and MacDonald, K.B. Soil and Ground Water
Pollution. *Proc., 12th Pacific Science Congress* $\underline{1}$,
p. 27 (1971).

69

Fried, J., Garnier, J.L., and Urgemach, P. <u>A Quantitative Study of Pollution of Ground Water: The Salt Content of the Water Table in the Haut-Rhin</u>. In *Qualite des Eaux, Pollutions, Deuxieme Partie,* French Bur. Réch. Géol. Minières, Bull. (Series 2), Section 3, No. 1, pp. 105-115 (1971).

+ + +

Fuhriman, D.K., and Baron, J.R. <u>Ground Water Pollution in Arizona, California, Nevada and Utah</u>. USEPA, WPCRS No. 16060 ERU, 253 pp. (1971).

Ground water contamination, dissolved solids, mineralization.

+ + +

Gelhar, --. <u>The Aqueous Underground</u>. *Tech. Review* $\underline{74}$(5), pp. 45-53 (1972).

Ground water contamination, New York City, Long Island.

+ + +

Gillham, R.W., and Webber, L.R. <u>Ground Water Contamination</u>. *J. Water Pollution Control Federation* $\underline{106}$, No. 5, pp. 54-57 (1968).

+ + +

Grossman, L.G. <u>Waterborne Strene in a Crystalline Bedrock Aquifer in the Gales Feny Area, Ledyard, Southeastern Connecticut</u>. USGS Professional Paper 700-B, pp. 203-209 (1970).

+ + +

Hall, G.M. <u>Ground Water in the Ordivician Rocks near Woodstock, Virginia</u>. USGS Water Supply Paper 596, pp. 45-66 (1964).

+ + +

Holly, D.E. <u>Transport of Dissolved Chemical Contaminants in Ground Water Systems in Nevada Test Site</u>. *Geol. Soc. Am. Memoires* $\underline{110}$, pp. 171-183 (1968).

Huling, E.E., and Hollocher, T.C. Ground Water Contamination by Road Salt: Steady State Concentrations in East Central Massachusetts. *Science* 176, No. 4032, pp. 288-290 (1972).

+ + +

Hutchinson, F.E. Environmental Pollution from Highway Deicing Compounds. *J. Sci. and Water Conservation* 125, No. 4, pp. 144-146 (1970).

+ + +

Jaag, O. Deterioration of Water and the Struggle Against Pollution of Lakes, Running Water, and Ground Water in Switzerland. *Soc. Sci. Nat. Phys. Maroc. C.R.* 37, pp. 15-29 (1971).

+ + +

Larson, T.E. Ground Water Supplies of Northeastern Illinois, Quality Problems with Well Waters. *J. Am. Water Works Assoc.* 56, No. 2, pp. 169-172 (1964).

+ + +

McCallie, S.W. A Preliminary Report on the Underground Water of Georgia. Georgia Geological Survey Bull. 15, 370 pp. (1908).

Contains a chapter entitled "Experiment Relation to Problem of Well Contamination at Quitman."

+ + +

McCracken, R.A., and Nickersas, H.D. Contaminants and Their Effects on Ground Water Supplies. *Sanitalk* 11, No. 1, pp. 11-27 (1963).

+ + +

McGaughey, P.H. Man-Made Contamination Hazards. In *Man and His Physical Environment,* G.D. McKenzie and R.O. Utgurd, eds. (Minneapolis, Minn: Burgess Publishing Co.) pp. 169-173 (1972).

Parizek, R.R.   Impact of Highways on the Hydrogeologic Environment . *Environmental Geomorphology* 1, pp. 151-200 (1970).

+ + +

Service Géologique d'Alsace et de Lorraine. Pollution of Ground Water, A Bibliographic Study. (Strasbourg), 2 vols., 65 and 179 pp. (1969-1970).

A two-volume report by the staff of the Geological Service of Alsace-Lorraine. The first volume considers mechanisms of infiltration of pollution. The second deals with pollution by detergents, pesticides, radioactive substances, and organic and inorganic compounds.

+ + +

Simmonds, M.A.   Carbon Dioxide in Domestic Water Supplies. *Proc., Soc. Water Treatment Exam.* 12, pp. 197-223 (1964).

Four principal types of water used as supply in Queensland: flowing surface water, impounding surface water, shallow ground water, and artesian ground water.

+ + +

Stanley, W.E., and Eliassen, R.   Status of Knowledge of Ground Water Contaminants. Federal Housing Administration Publication, Technical Studies Program, 465 pp. (1961).

+ + +

Sniegocki, R.T.   Ground Water Recharge--Natural and Artificial. (In *Ground Water Contamination, Proc. of the 1961 Symposium*), USDHEW, PHS, Robert A. Taft San. Eng. Cent., Tech. Rept. W61-5, pp. 16-20 (1961).

+ + +

Taylor, F.B.   Trace Elements and Compounds in Waters. *J. Am. Water Works Assoc.* 63, pp. 728-733 (1971).

Gives limits and observed concentrations for USA.

Tenu, A. The Pollution of Ground Water. *Hydrotechnica*
13, pp. 84-91 (1968).

Emphasis on karst.

+ + +

Walker, W.H. Illinois Ground Water Pollution. *J. Am.
Water Works Assoc.* 61, No. 1, pp. 31-40 (1969).

Case histories of pollution of ground water in Illinois.

+ + +

Walker, W.H. Salt Piling--A Source of Water Supply Pollution. *Water Supply Engineering* (July-August), Illinois
Water Survey, Urbana, Reprint Series 162, pp. 29-34 (1970).

Example from Peoria, Illinois.

+ + +

Walton, G. Problems in Ground Water Pollution. (In *Ground
Water Contamination, Proc. of the 1961 Symposium*), USDHEW,
PHS, Robert A. Taft San. Eng. Cent., Tech. Rept. W61-5,
pp. 1-6 (1962).

+ + +

Williams, J.H. Can Ground Water Pollution be Avoided?
Ground Water 7, No. 2, pp. 21-33 (1969).

South Missouri; general discussion of polluted areas.

+ + +

Witmoth, B.M. Salty Ground Water and Meteoric Flushing of
Contaminated Aquifers in West Virginia. NGWQS, Proc.,
Ground Water 10, No. 1, pp. 99-106 (1972).

+ + +

Wormald, B. Dorris-Butte Valley Water Quality Investigation. California Dept. Water Resources, Office Rept.,
23 pp. (1968).

12. SOLID WASTE DISPOSAL

Andersen, J.R., and Dornbush, J.N. Influence of Sanitary
Landfill on Ground Water Quality. *J. Am. Water Works
Assoc.* 59, pp. 457-470 (1967).

Eleven wells were installed at five locations around a
disposal area at Brookings, South Dakota and were sampled
at 3-month intervals.

+ + +

Andersen, J.R., and Dornbush, J.N. Quality Changes of
Shallow Ground Water Resulting from Refuse Disposal at
a Gravel Pit. Technical Completion Report, Water
Resources Institute, Washington, D.C. (OWRR Project
A-003 South Dakota), 41 pp. (1968).

An evaluation of the effects of landfill on the ground
water quality of Brookings, South Dakota.

+ + +

Apgar, M.A., and Langmiur, D. Ground Water Pollution
Potential of a Landfill Above the Water Table. NGWQS,
Proc., USEPA No. 16060 GRB 08/71, pp. 76-96 (1971).

Properties and movement of landfill leachate through
unsaturated soil (Pennsylvania).

Bergstrom, R.E. Feasibility Criteria for Subsurface Waste
Disposal in Illinois. *Ground Water* <u>6</u>, No. 5, pp. 5-9
(1968b).

+ + +

Bergstrom, R.E. Hydrologic Studies, Key to Safety in
Waste Management Programs. *Water and Sewage Works* <u>116</u>,
pp. 149-154 (1969).

Sources of contamination in Illinois ground water;
disposal of liquid contaminants.

+ + +

Bergstrom, R.E. Disposal of Wastes--Examples from Illinois.
In *Man and His Physical Environment,* G.D. McKenzie and
R.O. Utgurd, eds. (Minneapolis, Minn: Burgess Publishing
Co.), pp. 165-168 (1972).

+ + +

Bishop, W.D., Carter, R.C., and Ludwig, H.F. Water Pollution Hazards from Refuse-Produced Carbon Dioxide. *Proc.
3rd Int. Conf., Advances in Water Pollution Research
(Munich),* <u>1</u>, pp. 207-228 (1966).

+ + +

Blewer, N.K. Geology Considerations in Planning Solid
Waste Disposal Sites in Indiana. Indiana Geological
Survey, Environmental Study No. 1, Spec. Rept. 5, p. 7
(1970).

Factors: native of ground water supplies, thickness
of glacial shift over bedrock aquifers.

+ + +

Bucksteeg, W. Garbage Disposal and Its Effect on Surface
and Ground Water. *Gas-u. Wasserfach.* <u>110</u>, pp. 529-537
(1969).

Details, case history (in German).

Calvert, C.K.   Contamination of Ground Water by Impounded Garbage Waste.   *J. Am. Water Works Assoc.* 24, p. 266 (1932).

+ + +

Cartwright, K., and McComas, M.R.   Geophysical Surveys in the Vicinity of Sanitary Landfills in Northeastern Illinois.   *Ground Water* 6, No. 5, pp. 23-30 (1968).

Chemically altered ground water is traceable in uniform earth materials.

+ + +

Cartwright, K., and Sherman, F.B.   Evaluating Sanitary Landfill Sites in Illinois.   Illinois Geological Survey, Env. Geol. Note 27, p. 15 (1969).

+ + +

Cartwright, K., and Sherman, F.B.   Ground Water and Engineering Geology in Siting of Sanitary Landfills.   *AIME, Trans.* 250, No. 1, pp. 1-6 (1971); *Abstr. Mining Eng.* 21, No. 12, p. 58 (1971).

+ + +

DuMontelle, P.B.   Geologic Investigation of the Site for an Environmental Pollution Study.   Illinois Geological Survey, Env. Geol. Note 31, 19 pp. (1970).

Creation of an environmental pollution study area, University of Illinois campus at Champaign.

+ + +

Farkasdi, G., Golwer, A., Knoll, A., Mattess, K.H., and Schneider, W.   Investigations on the Microbiology and Hygiene of Ground Water Pollution Downstream from Waste Deposits.   *Stadtehygiene* 20, pp. 25-31 (1969).

High counts on soil samples downstream at depth of 3-5 meters; water same depth, counts not low.

Feige, W. <u>Are Bogs Suitable Sites for Waste Disposal?</u>
Deut. Geol. Ges., Z. <u>122</u>, pp. 71-80 (1971).

+ + +

Flawn, P.T., Turk, L.J., and Leach, C.H. <u>Geological Con-</u>
<u>siderations in Disposal of Solid Municipal Wastes in</u>
<u>Texas.</u> Texas University Bureau Econ. Geol., Geol. Circ.
No. 70-2, 22 pp. (1970); in *Man and His Physical Environ-*
ment, G.D. McKenzie and R.O. Utgurd, eds. (Minneapolis,
Minn: Burgess Publishing Co.), pp. 111-116 (1972).

Favorable and unfavorable conditions for sanitary landfills; discussion and examples.

+ + +

Foose, R.M. <u>From Town Dump to Sanitary Landfill: A Case</u>
<u>History.</u> *Assoc. Eng. Geol. Bull.* <u>9</u>, No. 1, pp. 1-16
(1972); *Abstr. Ann. Mett., Prog. Abstr.*, No. 14, pp. 24-
25 (1972).

Polluted ground water experimentally cleaned up; recircu-
lation through weathered shale in Hershey, Pennsylvania.

+ + +

Freeze, R.A. <u>Subsurface Hydrology at Waste Disposal Sites.</u>
IBM J. of Research and Development <u>16</u>, No. 2, pp. 117-
129 (1972).

Computer-generalized flownets for selected site characteristics suggest travel time of pollutants ranging from
1 to 660 years.

+ + +

Gross, D.L. <u>Evaluation of Dekalb County Area Solid Waste</u>
<u>Disposal and General Construction.</u> In *Man and His*
Physical Environment, G.D. McKenzie and R.O. Utgurd, eds.
(Minneapolis, Minn: Burgess Publishing Co.), pp. 294-
299 (1972).

+ + +

Harvey, E.J., and Skelton, J. <u>Hydrologic Study of a Waste</u>
<u>Disposal Problem in a Karst Area at Springfield, Missouri.</u>
USGS Professional Paper 600-C, pp. 217-220 (1968).

Hughes, G.M., Farvolden, R.N., and Landon, R.A. Hydro-
geology and Water Quality at a Solid Waste Disposal Site.
Proc., 7th Annual Eng. Geol. and Soils Symposium, Idaho
Department of Highways, pp. 116-130 (1969).

Northeast Illinois; leachate assimilated by environment
with and without significantly affecting ground water
resources.

+ + +

Hughes, G.M., Landon, R.A., and Farvolden, R.N. Summary
of Findings on Solid Waste Disposal Sites in Northeastern
Illinois. Illinois Geological Survey, Env. Geol. Note
No. 45, 25 pp. (1971).

Large areas of Northeastern Illinois could be used for
solid waste disposal with or without affecting ground
water.

+ + +

Hughes, G.M. Hydrogeologic Consideration in the Siting
and Design of Landfills. Illinois Geological Survey,
Env. Geol. Note No. 51, 22 pp. (1972).

Concepts based on conditions in Northeastern Illinois.

+ + +

Hughes, G.M., Trembley, J.J., Anger, H., and D'Cruz, J.
Pollution of Ground Water Due to Municipal Dumps. Dept.
Eng. Mines Resources, Inland Water Branch, Ottawa,
Tech. Bull. 42, 102 pp. (1972).

+ + +

Hughes, G.M., Trembley, J.J., Anger, H., D'Cruz, J. Pollution of Ground Water Due to Municipal Dumps. *Water and
Pollution Control* 110, No. 1, pp. 15-17 (1972).

+ + +

Kaufmann, R.F. Hydrogeology of Solid Waste Disposal Sites
in Madison, Wisconsin. University of Wisconsin (Madison),
Water Resources Center, Tech. Rept., 361 pp. (1970).

Kazman, R.G.   Exotic Uses of Aquifers.   *Proc., Am. Soc.*
Civil Engineers, J. Irrig. Drain Div. 97, No. IR3,
pp. 515-522 (1971).

Effects of waste disposal, gas storage, and other exotic
uses of aquifers.

+ + +

Keeley, J.W.   Solid Waste, Its Ground Water Potential.
NGWQS, Proc., USEPA, No. 16060 GRB 08/71, pp. 121-135
(1971).

+ + +

Landon, R.A.   Application of Hydrogeology to the Selection
of Refuse Disposal Sites.   *Ground Water* 7, No. 6, pp. 9-13
(1969).

+ + +

Langmiur, D., Parizck, R.R., and Apgar, M.A.   The Chemical
Interaction of Sanitary Landfill Leachates with Unsaturated
Soil in a Carbonate Rock Terrain.   Pennsylvania Dept.
Health, Progress Report, 27 pp. (1970).

+ + +

Lessing, P., and Reppert, R.S.   Geological Considerations
of Sanitary Landfill Site Evaluations.   West Virginia
Geological Survey, Bull. No. 1, 33 pp. (1971).

+ + +

McMaster, W.M., Betson, R.P., and Ardlis, C.V.   Some
Problems Associated with the Hydrology of Carbonate
Terrane.   *Geol. Soc. Am., Abs. with Prog.* 5, No. 5,
p. 41 (1973).

+ + +

Mollweide, H.U.   Some Observations on the Paper by Schraeber,
Ashes and Garbage Dumps and Their Effects on Ground Water.
Z. Angew. Geol. 17, No. 12, pp. 536-537 (1971).

Noring, F., Farkasdi, G., Galwer, A., Knoll, K.H.,
Matthess, G., and Schneider, --. The Decomposition Pro-
cesses in Ground Water Pollution Downstream from Waste
Deposits. *Gas-u. Wasserfach.* 109, pp. 137-142 (1968).

Brief review of existing literature on the decomposition
processes involved in ground water pollution. Detailed
account of investigations on two dumping sites in
Hesse, Germany.

+ + +

Otton, E.G. Solid Waste Disposal in the Geohydrologic
Environment of Maryland. Maryland Geological Survey,
Rept. Inv., No. 18, 51 pp. (1972).

Five terrane types for waste disposal.

+ + +

Riccio, J.F., and Hyde, L.W. Hydrogeology of Sanitary
Landfill Sites in Alabama: A Preliminary Appraisal.
Alabama Geological Survey, Circ. No. 71, 23 pp. (1971).

+ + +

Roessler, B. Influence of Garbage and Rubbish Dumps on
Ground Water (Translated by R. Zehnpfennig). *Vom Vasser*
18, p. 43 (1950).

+ + +

Schnieder, W.J. Hydrologic Implications of Solid Waste
Disposal. USGS Circ. 601-F, pp. 1-10 (1970).

Extent of pollution largely dependent on geologic
environment.

+ + +

Society for Water Treatment and Examination. The Effect
of Tipped Domestic Refuse on Ground Water Quality. *J.
Soc. Water Treatment Exam.* 18, pp. 15-69 (1969).

1. T. Walerton discusses details of garbage tips in the
   area of the Sunderland and South Shields Water Co.
   and their effects on well waters in the vicinity of
   the tips.

2. R.D. MacLean gives the results of a survey completed in 1963 of the water quality of wells and boreholes in North Kent.

3. A.S. Davison confirms the potential risk to ground water supplies from highly polluted effluents from garbage tips.

4. W.S. Holden provides tables of garbage tip location and boreholes and wells in the area of the Richmansworth and U-bridge Valley Water Co.

+ + +

State of California Water Pollution Control Board. <u>Effects of Refuse Dumps on Ground Water Quality</u>. Publication No. 24, 101 pp. (1961).

+ + +

Stewart, J.W., and Hanan, R.V. <u>Hydrologic Factors Affecting the Utilization of Land for Sanitary Landfills in Northern Hillsborough County, Florida</u>. Florida Bur. Geol., Map Series, No. 39 (1970).

+ + +

Stundl, K. <u>The Processes of Decomposition in Soil and Their Effect on Ground Water Quality</u>. *Gas Wass. Warme* <u>22</u>, pp. 142-147 (1968).

+ + +

Thompson, R.D., (ed). <u>Environmental Geology in the Pittsburgh Area: A Guidebook Prepared for the 1971 Annual Meeting of the Geological Society of America</u>. GSA Annual Mtg. Comm., 47 pp. (1971).

Landfill pollution, waste disposal.

+ + +

Tuck, L.J. <u>Disposal of Solid Wastes--Acceptable Practice A Geologic Nightmare</u>. Am. Geol. Inst. (Washington, D.C.), AGI Short Course Lecture Notes, Milwaukee, Wisconsin, 42 pp. (1970).

Warrick, L.F., and Tully, E.J.   Pollution of Abandoned
Well Causes Fond Du Lac Epidemic. *Engineering News
Record* 104, No. 10, 104 pp. (1930).

+ + +

Weaver, L.   Refuse Disposal, Its Significance. *Ground
Water* 2, No. 1, pp. 26-30 (1964).

Case histories from California and North Dakota.

+ + +

Williams, R.E., and Wallace, A.T.   Hydrogeological Aspects
of the Selection of Refuse Disposal Sites in Idaho.
Idaho Bureau Mines and Geology, Pamphlet 145, 31 pp.
(1970).

Disposal site selection in Idaho.

+ + +

Wolfskehl, O., and Boye, E.   The Effect of Deposited Ash
and Garbage on Ground Water. *Gas-u. Wasserfach.* 107,
pp. 36-38 (1966).

Laboratory experiments.

+ + +

Young, --.   Effects on Ground Water. *J. Water Pollution
Control Federation* 44, No. 6, pp. 1208-1211 (1972).

Ground water, water pollution, waste disposal, literature
review.

+ + +

Zanoni, A.E.   Ground Water Pollution from Sanitary Land-
fills and Refuse Dump Grounds, An Agricultural Review.
Wisconsin Dept. of Natural Resources, Research Report
No. 69, 43 pp. (1971).

+ + +

Zanoni, A.E.   Ground Water Pollution and Sanitary Land-
fills--Agricultural Review. *Ground Water* 10, No. 1,
pp. 3-16 (1972).